Kryon

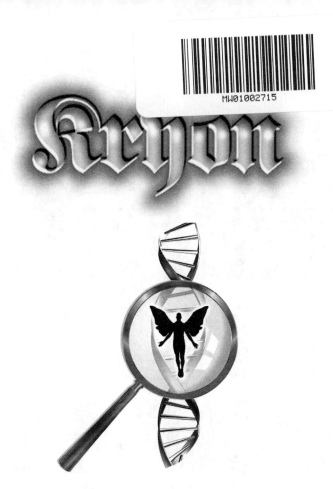

THE TWELVE LAYERS OF DNA

An Esoteric Study of the Mastery Within

Kryon
Book 12

182 International Kryon Books
in 23 Foreign Languages (not including Indigo book 3)

Spanish - 8
Kryon Books - One, Two, Three, The Parables, The Journey Home, Kryon Book Six, Seven, and eight

Spanish - 5
Kryon Books Eight, Nine, Ten, Eleven and The Indigo Children books

Hebrew - 3
Kryon Books - One, Two, Three

Hebrew - 10
Kryon Books - The Parables of Kryon, The Journey Home, Books Six, Seven, Eight, Nine, Ten, Eleven, and The Indigo Children 1 and 2

Slovene - 1
Indigo Children

Italian - 7
Kryon Books - One, Parables, Indigo Children, Book Seven, Eight, Ten, and Eleven

Estonian - 6
Kryon Books - One, Two, Seven, Parables + 2 Indigos

Turkish - 7
Kryon Books - One, Two, Three, Six, Seven, Eight, Nine and Ten

Turkish - 5
Kryon - Parables, Journey Home, Indigo Children & Book Eleven

Chinese - 4
Kryon Books - One, Two and Three & Parables

Portuguese - 7
Kryon Books - Indigo 1, Books One, Two, Three, Five, Eleven, and Indigo 3

Latvian - 7
Kryon Books - Seven, Eight, Nine, Ten, Eleven, Parables, Journey Home

182 International Kryon Books
in 23 Foreign Languages (not including Indigo book 3)

French - 12
Kryon Books - One, Two, Three, Journey Home, Six, Seven, and Eight, Nine, Ten, The Indigo Children books, Kryon book 11

Russian - 4
Indigo One and Two - Books 4 and 5

Russian - 13 (some not shown)
Kryon Books - One, Two, Three, Six, Seven, Eight, Nine, Ten, and Eleven

Danish - 1
Kryon Book One

Japanese - 3
Kryon Book - One, Five & Indigo Children

Hungarian 13
Kryon Books - One, Two, & The Indigo Children, Books Eight, Nine, Ten, Eleven, and Twelve

Bulgarian - 11 (not shown)
Kryon Books - One, Two, Four, Six, Seven and Eight

German - 13
Kryon Books - One, Two, Three, Journey Home, Parables, Six, Seven, Eight, Nine, Ten, Eleven, and Indigo 1 & 2

Romanian - 1
Indigo Children

Finnish - 10
Kryon Books - One, Two, Three, Four, Five, Six, Seven, Eight, Nine, and the Indigo Children

Greek - 4
Kryon Books - One, Two, Three, and Five (The Journey Home)

Greek - 4
Kryon Books - Six, Seven, Eight, and the Parables of Kryon

Dutch - 3
Kryon Books - Journey Home - Parables of Kryon - Indigo Children

Latvian - 6
Kryon Books - One, Two, Three, Six + Indigo 1 and 2

Korean - 1
The Indigo Children

Indonesian - 2
The Indigo Children

Lithuanian - 4
Kryon Book One and Five

DNA

noun (Biochemistry)
deoxyribonucleic acid, a self-replicating material present in nearly all living organisms as the main constituent of chromosomes. It is the carrier of genetic information.

noun (Quantum Biology - Metaphysics)
deoxyribonucleic acid, the double helix made of less than 5% biological instructions as the carrier of the genetic makeup of the Human body. More than 90% is quantum energy and instructions, which define sacred life and set up the Akashic history and the divinity within a Human Being.

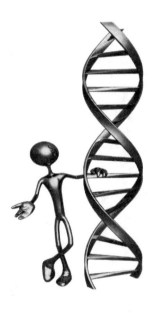

Dedicated to...

Patti Carroll

My loving wife of many years who behind the scenes personally organized more than 50 Kryon meetings a year for a decade! A true dedicated Lightworker for Kryon. You are in my Akashic DNA!

In memory of...

Tommy Thomsen

My loving German friend and companion who left us way too soon...

Ted Dircz

No Kryon Cruise will be the same without him!
So pour the wine, Ted.
We miss you so much!

Mary Lou Jackson

Our DC friend and profound teacher of children.
Always with a smile.
You will be missed!

THE TWELVE LAYERS OF DNA
An Esoteric Study of the Mastery Within
Kryon Book 12

Publisher: **Platinum Publishing House**

P.O. Box 4357
Sedona
AZ 86340

Kryon books can be purchased in retail stores, by phone or on the Internet at [www.kryon.com/store]. (800) 352-6657 - E-mail <kryonbooks@kryon.com>

The color illustrations in this book can be purchased on the Internet at [www.ElanDubroCohen.com].

Written by Lee Carroll
Cover Design by Elan Dubro-Cohen
Color Illustrations by Elan Dubro-Cohen
Editing by Dawne Brooks
Copyright © 2010—Platinum Publishing House
Printed in the United States of America
First Edition—First Printing—September 2010
Second Printing—November 2010

All rights reserved. Portions of this book may be freely quoted or reprinted up to 500 words without permission, provided credit is given to Platinum Publishing House—THE TWELVE LAYERS OF DNA, Copyright © 2010.

The Kryon® logo is a registered trademark with the United States Patent and Trademark office.

Table of Contents

Foreword by Dr. Todd Ovokaitys 9

Introduction - Lee Carroll 19

CHAPTER ONE
Invisible Stuff! 26
    Lee Carroll

CHAPTER TWO
How It Began 30
    Lee Carroll

CHAPTER THREE
DNA Examined in a New Way 42
    Lee Carroll

CHAPTER FOUR
Things That Nobody Thinks About 50
    Lee Carroll

CHAPTER FIVE
The Teaching Begins 70
    Lee Carroll

CHAPTER SIX
The Twelve Layers as Given by Kryon 100
    Kryon

CHAPTER SEVEN
DNA Group One - Layers One, Two, and Three 110
    Kryon

CHAPTER EIGHT
DNA Group Two - Layers Four, Five, and Six 134
    Kryon

CHAPTER NINE
DNA Group Three - Layers Seven, Eight, and Nine 152
    Kryon

Table of Contents - continued...

CHAPTER TEN
DNA Group Four - Layers Ten, Eleven, and Twelve 186
   Kryon

CHAPTER ELEVEN
The DNA Grouping Summary and their Big Scary Secret! 214
   Kryon

CHAPTER TWELVE
Activating the DNA Field 222
   Kryon in Boulder, Colorado - Live Channelling

CHAPTER THIRTEEN
Beginning to Activate the Specific Energies of DNA 236
   Kryon in Riga, Latvia - Live Channelling

CHAPTER FOURTEEN
The History of DNA and the Human Race 258
   Kryon in Portland, Oregon - Live Channelling

CHAPTER FIFTEEN
The Great Scientific Bias 280
   Kryon in Gaithersburg, Maryland - Live Channelling

CHAPTER SIXTEEN
Current Events - Transplants - Artificial Life - The Last Word 300
   Lee Carroll

KRYON INFORMATION
Web Site - The United Nations Channellings - Small Meetings ... 313

BASIC INDEX 318

THE DNA ILLUSTRATIONS
Full Color Illustrations of Each of the DNA Energies 321
   Elan Dubro-Cohen

Foreword
Kryon Book 12

Dr. Todd Ovokaitys

Foreword
Kryon Book 12
by Dr. Todd Ovokaitys

Ever since Watson and Crick revealed the double helical structure of DNA, this elegant image has been a cultural icon, reflecting our ability to peer deeply into the structure of our makeup and find beauty and form translating into the functions that produce life itself. Much as an image of Einstein conjures the notion of 20th century brilliance, the image of DNA resonates at a level of both aesthetic appreciation and having discovered intimate truths about our very nature as living, sentient beings.

Following the landmark discovery of the chemical form and structure of DNA, work proceeded to unravel the very code of DNA. In a few short decades, the triplet code of DNA had been fully deciphered, another testament to human scientific ingenuity. The triplet code means that every three bases of DNA translate to a specific instruction to start to make a protein, to add a particular amino acid building block, or to stop the process of making the new molecule. At the level of discovery, it was felt science had cracked the code almost fully and was ready to transform the sciences of health and biochemistry.

While the basic code was understood, the vast length of the code even in a small organism made the next daunting task the ability to map out the code of a complete living form. Two scientists at Johns Hopkins University Medical School then discovered the tool to make this feasible. Doctors Hamilton O. Smith and Daniel Nathans performed landmark research into a class of DNA enzymes called restriction endonucleases. In short, these enzymes were like very specific DNA clippers that would split DNA only

Foreword— Dr. Todd Ovokaitys

if a particular sequence of DNA bases was present. By being able to cleave DNA systematically in particular places and then overlap these smaller, more manageable sequences, the possibility of determining the entire DNA sequence code of a particular living organism became possible.

This aspect of DNA scientific discovery was especially poignant for me because I was a medical student at Johns Hopkins when the Nobel Prize was awarded for that work. Not only was I present, I was in the microbiology lab section that was being taught by Dr. Nathans at the time. He seemed like such a quiet, mild mannered person. I was surprised and pleased to learn that this unassuming lab instructor had just, by the way, won a Nobel Prize for a monumental DNA breakthrough. Not only that, instead of having to plate random microbes on a Petri dish to culture them that day, I got to drink champagne with my instructor and classmates, celebrating him and his colleague and the march of science. This was the academic equivalent of a snow day.

When this discovery was combined with the revelations of a San Diego surfing scientist named Kerry Mullis, the stage was set to pursue the most ambitious challenge in the history of DNA research—that of determining the sequence of the entire human DNA code. Kerry Mullis conceived of a method called the polymerase chain reaction, for which he was awarded a Nobel Prize. Without going into the details of how it works, this is an ingenious method that allows multiplying specific DNA sequences up to millions of times. Combining these and further technical evolutions, particularly in the automation of sequencing, the world embarked on the Human Genome Project. Supported by governments, universities and private industry, a few years ago this monumental task was completely achieved. For the first time in human history, we knew the exact recipe of DNA information that produced a Human Being. Three billion base pairs converted to letters would generate a textbook approximately 300,000 pages long! Such incredible complexity and detail had been completely deciphered and decoded.

From this monumental point of achievement, it was felt that medical science had the key for the healing of many diseases, and even learning the codes to extend our longevity to levels previously not thought possible. Within this blessing, there arose the astonishing discovery that it required fewer coding sequences called genes to make a human than was previously believed. It had been thought that the Human Genome, the total DNA code of the 23 chromosome pairs of the human genetic code, would harbor at least 100,000 genes. It was somewhat astonishing when it was learned that there were only 30,000 to 40,000 genes required to code for a person. Even more astounding, if not shocking, was the discovery that it appeared that only a small fraction of the Human Genome had base sequences that were actually coding for genes. This seemed like spacers in a recording tape that had no information between the actual coding sequences. The perplexing finding was that this noncoding DNA, initially labeled as "junk DNA," comprised 97% of the entire human code!

Why was so much of the system noncoding? If it did not have a particular purpose, how had it survived evolutionary pressures that rarely preserve features not of definite benefit to the organism? The high proportion of noncoding DNA suggests it is somehow important in the orchestration of what the coding sequences of genes do. While we have the recipes for building the 30,000 to 40,000 key proteins that produce human life, it is much like having a recipe with thousands of ingredients but not the instructions for how to create the confection. While the mystery of the "junk DNA" continues to be sorted through, one of the explanations is that these patterns are related to how the ingredients of the recipe are put together as regulators of the outplay of the overall code.

Foreword— Dr. Todd Ovokaitys

Before the extraordinary promise of advances in treatment of medical conditions, or overcoming hereditary predispositions or enhancing the quality of being human, the complexities of regulation of the code will need to be more fully understood. While there have been many advances in the understanding of particular regulatory systems, the full unveiling of the inner workings of regulation will likely require fuller knowledge of the actions and roles of the non-coding DNA elements, the vast majority of the genomic multi-layer cake.

In this context of a brief tour of the startling advances in understanding DNA and the genetic code in the last six decades, it also opens the door to the mystery of the deeper functions of how the system works. Often in the history of science, it is an intuitive leap that leads the way to discovery—much as Kerry Mullis' vision of DNA gave him the insight that allowed him to discover the polymerase chain reaction that revolutionized genomic science.

The invitation to offer introductory thoughts for this book from Lee Carroll, internationally renowned author of best-selling self-development books, was to provide a bridge between the rigorous scientific world of DNA and the intuitive, improvable world of inner visions of DNA. While such visions may or may not make a contribution to the ultimate understanding of DNA both for knowledge and for practical use, many scientific breakthroughs have been heralded by a flash of insight followed by years of rigor to unveil deeper wisdom.

As a second-year undergraduate student pursuing chemistry studies, one usually learns a well known and oft repeated example of the intuitive leap that paved the way for organic chemistry to begin and then flourish as a field. This is the story of the chemist known as Kekule, who was among the researchers striving to determine the structure of the benzene molecule. With six atoms of carbon and six atoms of hydrogen, it did not fit the behavior expected of small organic molecules—organic in this sense meaning with a backbone of linked carbon atoms. One night in a dream, Kekule saw a snake swallowing its tail. This seemingly unrelated image inspired him to explore benzene being a ring molecule, as though it were a snake swallowing its own tail. This revelation allowed him to prove the ring structure of benzene with novel molecular bonding properties that paved the way for the advances of modern organic chemistry.

While Kekule is credited with the dream that he allowed to inspire him, the image revealed to him was ancient. Known as the Ourboros, ancient mystery school traditions conjured this image as a snake or dragon or lizard swallowing its own tail. This was felt to represent the great circle of existence with the tendency of patterns to recur in cycles and was well known in the Hermetic tradition of alchemical exploration.

If a process goes through an entire circle and returns to the beginning a little more developed than before, the pattern mapped out is a circle with an ascendancy of quality at each cyclic completion. Rather than a circle, this maps out a spiral. In a process of learning described as hermeneutic, returning to the same information after one has had intermediary learning allows greater learning and insight from the same information—a so-called hermeneutic spiral. The essential geometry of DNA is a spiral, its very form a suggestion of hermeneutic learning through cycles of growth and development.

The double helical nature of DNA is a spiral combined with its counter-opposed spiral. The double helix simultaneously spirals in both directions at the same time. This also is parallel to the oft repeated Hermetic assertion of "as above, so below."

Foreword— Dr. Todd Ovokaitys

The information offered by Lee Carroll in this esoteric book is a type of advanced Hermetic information put in terms that can be fathomed, unlike much of the coded mystery school traditions. None of this information may ever be provable, yet it may allow the ability to peer deeper into the core of DNA and see if it is willing to reveal any answers not before seen.

Perhaps the most profound development not in chemistry but in physics that may relate to the intuitive vision Lee Carroll is offering is the belief that our Universe has a higher dimensional ordering than the three dimensions we perceive. According to superstring theory, there are a handful of dimensional models that allow the mathematics for creating unified formulas of fundamental forces in physics. One of these is a 10-dimensional model in which the three dimensions of space have added unto them a dimension of time and six deeper dimensions that are unseen. According to this mathematics and theory, these six extra dimensions curl up into ultra-tiny balls that surround all the points in our four-dimensional space. Though unseen, these allow the full outplay of all the forces of nature.

Similarly, with the intuitive information Lee offers with the signature of Kryon, it is clearly stated that much of what is being described may never be provable with physical, scientific instruments. This is based on the notion that it is not possible to understand with clarity a higher dimensional object or phenomenon with a lower dimension instrument. It is not possible, for example, to understand fully the concept of sphere if your only dimension of measure is that of a circle.

Peering into the deep structure of reality, the book *Warped Passages: Unraveling the Mysteries of the Universe's Hidden Dimensions* goes beyond previous work for the nonspecialist reader in describing the latest evolutions of thought in higher dimensional physics. Written by Lisa Randall, a Harvard professor of physics, Warped Passages offers a dizzying view of the possible unseen dimensions that make up our physical reality. Beyond the ultra-tiny, six dimensional balls wrapped around points in 3D space, the latest theory extends to higher dimensions occupying membranes of a wide range of configurations. These could be long, curled up tubes or infinitely long but infinitesimally thin sheets or many other potential geometries. The mathematical spaces and dimensional constructs in the deep not directly detectable inner reality of our Universe are undergoing intense exploration. The inner world of our Universe may be as complex as our imagination and thought can reach.

In the context of the theory of our physical space being of a higher dimensional nature that at most may reveal its shadow enters a DNA experiment that inspires profound inquiry. Performed by a Russian physicist named Vladimir Poponin, he did a DNA study that gave unexpected results at several levels. In a study chamber the polarization and orientation states of light waves known as photons were measured. As anticipated, these waves of light moved randomly in the experimental chamber. He then placed DNA within the chamber and measured the photons again. To his surprise and beyond expectation, the presence of DNA had strongly organized the light waves into a coherent pattern, suggesting that DNA produced a profoundly powerful field that strongly organized the space around it. The presence and potency of this effect was vastly beyond what would have been predicted from mere chemical principles.

With a satisfying discovery in hand, Poponin then did the appropriate test to restore control conditions and complete the experiment. He removed the DNA and measured the photon characteristics again. The reasonable expectation is that without the physical DNA being present, the photon pattern should simply have reverted to a random ordering. Most astonishingly, when the physical DNA was removed, the photons remained in an organized pattern. This

Foreword— Dr. Todd Ovokaitys

would be much like removing a magnet that was organizing iron filings and have the filings remain in an organized magnetic flux field pattern. This suggested, without a physical theory established to describe why this would be so, that there remained a potent residual effect in the space simply because DNA had occupied that space. The implications for the profound effect of DNA to have a residual and lasting field and informational organizing effect on space requires new theories to account for this effect. If nothing else, the profundity of living DNA, human or otherwise, on the structuring of space has been demonstrated though not yet explained.

While there may arise physical theories that explain this so-called "phantom DNA effect" discovered by Poponin, this experiment raises the possibility that the nature of reality is well beyond that which meets the eye. It has been the pattern of 20th century physics that the more the attempt has been to prove that our reality is objective, experiments show that the effect of the observer is intimately related to the resultant reality that is observed.

The "phantom DNA effect" combined with the observation that 97% of the Human Genome is felt to be noncoding "junk" indicates that there are more layers of mysteries to unravel in our understanding of DNA, what it is, and how it truly works. Whether the intuitive information provided in this esoteric book offered by Lee Carroll in any way reflects the deeper reality of DNA, we may never know. That it reflects an intuitive leap that may inspire with its beauty or its elegance may be its greatest contribution. And as such, it may either contain its own elements of truth, or may some day contribute to a flash of insight that takes us another step forward in revealing the mysteries of this coding element of life.

Dr. Todd Ovokaitys

The Twelve Layers of DNA

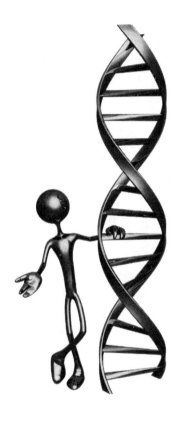

Introduction

Lee Carroll

Kryon
Book 12

Introduction
Kryon Book 12
by Lee Carroll

I am Lee Carroll, the original channel for Kryon, and an author and lecturer in the New Age. I tell you this since I'm assuming that you picked up this book well aware that it is esoteric information channeled by a beautiful energy from beyond the veil. That means that almost everything to follow is from a source that is metaphysical in nature, and not a scientific study.

I travel the earth channelling Kryon to thousands of attendees in auditoriums in very exotic places—Moscow, Jerusalem, Santiago, Caracas, Paris, Geneva, Bogota, Riga, Sao Paulo; my work takes me to areas that I thought I would never see. For 21 years and with 14 other books, I have expounded on the messages of Kryon, now in 24 languages worldwide. Since 1989, I have been delivering messages of self-empowerment and spiritual guidance through this loving entity, and have been invited and channelled seven times at the Society for Enlightenment and Transformation at the United Nations in New York City. Never in my work, however, has there been a single subject that would occupy such a long ramp-up in learning, or an entire book, until now.

In 2003, Kryon began giving me indications that something special would be revealed that would create a book unto itself. At that point, I never thought I would be required to learn Hebrew words or study numerology to present it. But this obviously was part of the plan. Kryon began giving me the full esoteric meanings behind the DNA energy within our bodies. He did it carefully over a long period of time so that it would stick within my memory and be written down and studied.

Introduction

There are a total of 12 energies or layers of study around our DNA, and each one has a Hebrew name. In addition, each Hebrew name is a "name of God." To make it even more esoteric, each one also has a numerological energy and a complex interaction with the others in the group of 12. This means that instead of just a "list" of the DNA layers, I would have the revelation of a "system" before I was done... a system of sacredness and purpose, and one that knocks you over with an overall message of love and appropriateness.

You can't read this book and not be aware of your magnificent place in the Universe. When you are finished and put the book down, you have to take a deep breath and ponder... "If this is all true, then I really am part of a Universal plan of God!" You would be right in this proclamation, for this is really what the book is about.

Eventually, each layer will be presented one by one, with the appropriate teaching around it. In order to really get something out of this book, you have to see all of this, however, not as a study of 12 things, but a study of the enormity of a system that somehow places "pieces of God" on the planet, and then asks them to discover "the rest of the story." That's us! Could it be true? If so, what will you do with the information?

Don't read this book if you just want some more things to add to your list of what you know. Instead, read it with the idea that there is a great deal more you are about to discover about yourself. Then also realize that with self-discovery comes change, often big change, and many more questions.

The first chapter after this introduction is a discussion of why this information and energy is invisible to the logic of our minds, and especially why it is that you cannot see it with your eyes. We are all beginning to "connect the dots" these days with what science is starting to also acknowledge. Science is not beginning to prove our esoteric points, but rather it is beginning to provide a scientific bed for our esoteric information to lay in, which gives us permission to move our previously "strange ideas" to the "possibility of reality" to the eyes of many. Simply put, what is happening is that there is a confluence of scientific fact that is beginning to fit with this very study. Spirituality and science may, indeed, someday be so similar that it will be hard to separate them.

After that I will give you the lineage of this study and discuss those who participated, willingly or not, so you can understand where some of the Hebrew and numerology is coming from and why things got arranged the way they did.

There is a discussion of the structure of DNA following that, and a revelation of things that are "out of logic" becoming far more logical. This includes the latest on "junk DNA," what it might be, and why it was ever called that.

My favorite discussion is entitled, "Things nobody every thinks about." There are things in front of our very faces that defy logical thinking, yet we ho hum our way through life, never even looking at the dichotomy they represent! I cover some very controversial and esoteric information that is the basis behind the Kryon teaching. You have to know a larger picture of what has happened on the earth before you can then launch into this study of what is in your DNA. For much of what Kryon lists in our DNA is historical and carries within it more than 100,000 years of "who you might be." This is a big chapter, divided into several parts. Don't miss the part about me in Australia, with the flies.

In the chapter called, "The Teaching," I'll speak of how Kryon wants us to see DNA, and how he wishes us to consider it. I'll also discuss basic numerology in that chapter, and it's a far more comprehensive topic than some may think. It's a multidimensional discussion and as complete as I can make it as an introduction to that science.

Introduction

Next, the discussion begins, layer by layer, of what Kryon speaks of regarding Human DNA—the names, the energies, the purposes, a blueprint of "mastery inside." The discussion has a twist at the end, as you will see, but I am not going to give it. Kryon is.

As you read this book, you will see that the actual definitions and teaching about the individual layers is not the actual majority of the book. This is on purpose, for the book does not exist to give you still another list of things to have in your metaphysical locker. Instead, I hopefully wish to surround you with the profundity of what DNA actually might be, and what it really means to us. Within the layer discussion, Kryon often opens up to other subjects, some stories, and even attributes of ET's! So it's a lively discussion to say the least.

Next I will present four channellings by Kryon about the activation of DNA, Human scientific bias, and the multidimensional existence in general. These channellings were given in the last months of writing this book and are an excellent way to bring the practical applications of this esoteric teaching into a larger focus.

Sometimes there are duplications in information within the book, and especially within the live chanellings. Just walk through it, for the live channellings are given independent of the information within this book, so there are several explanations of the "junk" DNA being the real energy of our spirituality. To edit this out diminishes the energy of the live experience. Anyway, I personally don't think we can hear it too much!

Be aware that all through this book, Kryon and I use the word *quantum* interchangeably with *multidimensional*. The real physics meaning of the word quantum refers to a minimum unit of matter or energy in a transfer process. The word comes from the Latin "quantus," or *how much*. However, we are using the word as a popular colloquialism, specific to what people are hearing today, instead of the actual physics terminology. Therefore, a "quantum state" means "a multidimensional state" in this book. Probably the reason this term took on this popular meaning was because most of physics refers to things in a pure, empirical state. However, the theory of quantum mechanics gave us the beginning of the understanding of things that seemed to be in a "random" state, or what some even call "chaos." This randomness or "probability-based reality" may only be the way we perceive it and not random at all by the standards of new physics laws we don't yet have or understand.

Lastly I give you some current events to ponder. What do you think Spirit's reaction is to Human organ transplants? If DNA is absolutely unique, and contains our Akashic record, what is the spiritual integrity of a transplant? Should it be done? What if the person is going to die without it? There are some very good common sense answers in this section.

Then I discuss the latest "creation of man-made life," the very small DNA molecule that has been totally created in the laboratory in 2010. Is it proper? Does it affect us? What are the ramifications of it all? Again, it's current and there is common sense presented within the answers... which always seems to have the energy of offending at least a few.

In the very back of the book are the beautiful full color channelled illustrations of the DNA layers by artist Elan Dubro-Cohen. These are truly a beautiful addition to this book!

The last remarks will be about the fact that this book will be copied and called something different, added to, reworked with another name, made into the evil manifesto of the century and other common things that all happen to my books.

But hey... without that, where would the fun be?

Chapter One

Invisible Stuff

Kryon

Book 12

Lee Carroll

Chapter One
Invisible Stuff!
by Lee Carroll

The foreword to this book was written by researcher and friend, Dr. Todd Ovokaitys. His comments in this preface do not mean he endorses all the esoteric information here. Instead, I asked this M.D. to write only about the possibilities of multidimensional, biological energies, since this is exactly what his profound DNA work has discovered—interdimensionality within our own cellular structure. His patents and subsequent work with world-class scientists in Russia are going to help heal the African continent. His work is viable, mainstream, and will help eliminate AIDS from the list of diseases you "can't cure." His work has been noticed, even by heads of state.

The premise of "invisible" things that we have before us every day was pegged in about 1999 with the establishment of the Super String Theory. Physics is evolving, and this was one of the biggest swings from standard physics to a far more universal quantum physics. You can see this today very clearly within the excitement of the newly opened atomic accelerator in Geneva, Switzerland. Most of the physicists who are on the "Atlas Experiment" are what are called "stringers" (their name, not mine). They are, indeed, Super String Theorists. Here is a description of what this is all about, from the source:

ATLAS is a particle physics experiment at the Large Hadron Collider [CERN]. Starting in Spring 2009, the ATLAS detector will search for new discoveries in the head-on collisions of protons of extraordinarily high energy. ATLAS will learn about the basic forces that have shaped our universe since the beginning of time and that will determine its fate. Among the possible unknowns are the origin of mass, extra dimensions of space, microscopic black holes, and evidence for dark matter candidates in the universe.

ATLAS brings experimental physics into new territory. Most exciting is the completely unknown surprise—new processes and particles that would change our understanding of energy and matter. [http://atlasexperiment.org]

One of the core statements of Superstring Theory is that "The center of the atom contains at least 11 dimensions." The Atlas experiment sets out to prove this in only the ways physicists will accept, through an exhaustive collection of data analyzed for years. It was this way at Fermi Lab with the discovery of the "top quark," a totally theoretical atomic particle that could only be proven with a device the size of a small city. Now science has a larger device (the size of a larger city), and they are going after the "Holy Grail" of physics, opening the door to a multidimensional world. It's exciting to men and women with white coats who look at squiggly lines on digital prints for hours. But eventually it will show what they already suspect—that you and I have a multidimensional world around us, yet we only are aware of four dimensions in an 11 or more dimensional soup. So what's in the other dimensions? Why can't we see evidence of them? Perhaps we can.

Dr. Ovokaitys, with the help of Scott Stratken, a physicist and co-inventor of the sonogram, created a device that somehow excites multidimensional parts of our cellular structure with information. The results of lab tests and even Human trials showed that the Human body was reacting to things that had never been seen before. Somehow "invisible biology" was being stimulated to create self-diagnostics, and in many cases, to attack virulent, marauding diseases to the point that they became undetectable. In other cases even stem cells were being "informed" to create new tissue.

In another set of experiments totally unrelated to Dr. Todd's work, but mentioned by him in his foreword, Dr. Vladamir Poponin discovered a multidimensional field around DNA itself! Dr. Poponin is a *quantum physicist* who is recognized worldwide as a leading expert in quantum biology, including the nonlinear dynamics of DNA and the interactions of weak electromagnetic fields with biological systems. He is the senior research scientist at the Institute of Biochemical Physics of the Russian Academy of Sciences and was on loan to an American research institution when these remarkable series of experiments were carried out.

Dr. Poponin discovered that DNA actually has a multidimensional field around it, and this field was powerful enough to change matter in a precise and controlled experiment implemented with a very small sample of living Human DNA and photons. The DNA consistently patterned a random light source particle array into the symmetry of a sign wave! This was completely unexpected and led to an odd name, but one that says it all: *Phantom DNA*. The "phantom" name was given to exemplify a totally invisible part of the energy of DNA that is always there (http://twm.co.nz/DNAPhantom.htm). It seemed that DNA was susceptible to receiving information from an outside source that actually gave it instructions.

So it appears that the idea of multidimensionality within cellular structure doesn't entirely belong to the New Age *"Fruit Loop and wind chimes crowd"* after all! Indeed, the New Age is often blamed for all kinds of wild and unsubstantiated theories that appear foolish and worthy of a heavy eye-rolling. In fact, the idea of DNA having multidimensional properties, including at least 12 energies, was one of those wild theories, until—oops—science started coming around to proving some of it. Now it appears that DNA is, indeed, multidimensional, and that means we are, too! The "protocol" bed is made, and the wild, unsubstantiated things have started to become believable, if not common.

Chapter Two

How it began...
Kryon
Book 12

Lee Carroll

Chapter Two
How it Began...
by Lee Carroll

I don't give you the lineage of this journey just to fill space in this book. It's important you know the sources I have used, and how some of this information came to be. This will enhance the message for you as it is revealed.

In channelling, many feel that the Human Being's consciousness is supplanted by another entity, and that during this time the Human has nothing to do with anything. It's a "takeover" and often is supernatural. This also gives the idea that somewhere, somehow, this is an occult experience and, therefore, is very strange and odd. In fact, all of this is incorrect.

I've been doing it successfully for 21 years. And I would like to tell you that it requires a total meld of everything the Human knows and believes, combined with a sacred permission to join together a consciousness from the other side of the veil with a Human. In effect, the channeller becomes a translator, who translates multidimensional messages into 3D. He/she is a "linearization specialist." If you want to see how difficult this is, then just take a look at the various psychics, readers and futurists who regularly try to do this, and end up delivering profound information about future events that never happen. There is far more here than anyone thinks is actually taking place, and perhaps someday I'll give more of an explanation of what it's really like. But each of us who channel have a very different kind of experience. This alone should give you a clue that it is personal to the path of the Human Being, and that there is no "generic channelling" method that all should follow. As Humans, we are a very singular and linear lot; however, there are still those who will point a finger and say, *"He's not doing it right."* Let me interpret that: *"He's not doing it my way."* There is no right or wrong, and there is no empirical manual on channelling. It's personal and private, and each person who does it knows how unique an experience it is.

It's the closest thing you can do to "get in touch with the God within," yet many feel that God (or what you think of as God) has nothing to do with it. Kryon has told us over and over that channelling is the method of transfer of all spiritual knowledge. He reminds us that all the Holy scriptures on the planet, no matter what your religion, were written by Humans. "God" wrote none of them! Therefore, there has to be some kind of acknowledgment of this bridge between the message and the messenger. If Humans wrote all this (which they did), then where did they get the information?

In Christianity, even the Pope has a method for receiving divine information and a sanctioned place he can be where everything he says is infallible. As Christ's Vicar, he has the right to sit in the Chair of St. Peter in the Vatican, where a process called "Ex Cathedra" takes place. This is channelling, but don't tell a Catholic that (please). No need to rock that boat.

Even Paul the Apostle, while writing to his friends from a jail cell in Rome, had this very attribute—for these letters to friends became at least seven books of the New Testament! Yet when Christians speak of the Bible, it is the "word of God." Actually, it is the word of a Human under the influence of spiritual energy. Paul "channelled" much of the New Testament via his letters. To many Christians, it was the "Holy Spirit" who came to Paul and gave him his information, which is used by Catholic and non-Catholic alike. In all cases, there was an unspoken and unseen energy that allowed a Human to speak about the truth of God. This process is accepted in every religion.

But today, according to the reactions of some religious leaders, God has stopped talking to us completely, for no "new" information is allowed. There is the unspoken rule that all scriptures that will ever be written have been written. Channelling is, therefore, fake. Does this make sense to any spiritual person reading this? Who passed the law that God stopped talking? *"Wait a minute, Lee!"* my church friends say. *"We never said that. God talks to us all the time! We just said that God doesn't talk to YOU!"* Oh... and that's because? *"You're not in our church."* Oh. I see.

My God is personal, alive and dwelling in my cellular structure! Messages are forthcoming all the time, and they don't disagree with others who are getting the same kind of energy. In fact, they don't even disagree with Paul! For I firmly believe that the messages of the ancients, no matter what religion, were all about self-empowerment and finding peace within ourselves and between others on this planet through finding the creator within. If you really go back and study the original messages from the greatest prophets on Earth, there is a wonderful symmetry of the messages and a beautiful sameness of thought.

The message I received from Kryon some years ago was as follows, *"You are going to be given information about the 12 layers of DNA. This information will be the first of its kind. Each layer will have a Hebrew phrase name, and will be a 'name of God' in ancient Hebrew. Before I start the information stream, go get some direction from the teacher, Barbra."*

So here is another premise that breaks the norm. To many, channelling is supposed to be completely and totally spontaneous, yet here I am getting information about what is coming—and getting instructions to get help! Suddenly, you may see a system here, where "the family" is involved in some of the higher-level information that is coming through the veil. In my case, this certainly was the case. Kryon wanted me to do some research before the channelled messages began. Why? Because it's the Human who must explain it, even if it's channelled. In my case, Kryon wanted me to get intuitive information from yet another channeller, to be part of the one he was going to give to me!

Enter Barbra Dillenger

The teacher, Barbra, is Dr. Barbra Dillenger, world-class reader and profound numerologist who lives in partial retirement in Del Mar, California, where my work began. Synchronicity has her right in my backyard, and so I got an appointment. There is kind of a private joke among those of us in spiritual work. There is always someone coming to us and saying, *"God told me you have a message for me. What is it?"* We get this all the time. The interesting part of this is that rarely are we in the loop, and God didn't say anything to us about "the message." So it becomes kind of a funny thing that we all go through, as we try to tell those who say this that timing is the key, and when we get the message, too, we'll contact them. It's the best we can do without making them feel bad. The real story here is that anyone with a message like this for us should realize that if God has a message for them, they can get it anytime they wish from their own "innate source." Gone are the days when you have to seek out an authority for you own, personal message. In fact, immediately popping up and running to a reader or channeller takes all the responsibility away from you and places it on us. Many understand this later on. Meanwhile, we get this constantly.

So here I am, in Barbra's office, telling her that, *"Kryon told me you had something for me."* (Sigh). I felt sheepish at best, and a little smile crept in as well. Would she burst out laughing? Well, she didn't. Instead, she smiled and went to her bookcase. There, she pulled out a beautiful out-of-print book by one of the world's finest spiritual historians, J. J. Hurtac, entitled *The Seventy-Two Names of God*. All of the names of God were in Hebrew. She looked at me and said, *"What you need is in here."*

Indeed, it was, for as soon as I opened the book, Kryon said to me, *"Of these 72 names, 12 of them are the names of DNA. Learn them all in Hebrew, so that when I use them in channelling, you will know how to pronounce them. The meanings I give you will be different than the ones in the book."* So I immediately faced an issue; I should know at least a little Hebrew pronunciation so I wouldn't butcher the names when they occurred later in channelling. This is where the second person would become involved, and in a way that was far larger than just helping me with Hebrew. I'll tell you about Elan Dubro-Cohen in just a moment.

Enter J. J. Hurtac

J. J. Hurtac has been teaching long before I ever came on the scene. My first experience, as for many others, was with his book *The Keys of Enoch*. This is one of the most profound metaphysical and scientific books in existence, and it was given by J. J. using his research and his intuition. It's so heady that most people can't understand much of it, but feel they are getting it anyway! Some just give up and say, *"There is wonderful information in here, so I'll put it under my pillow and hope I get it as I sleep!"* You laugh? Go get the book and see if I'm right.

Hurtac does not see himself as a channeller, but as a spiritual historian. It's tough to agree with this as you read *The Keys*, since so much is verbatim from a source that is not Hurtac. But I wanted to honor him in these pages, so I'll call him a historian. The information he has continually provided has been timely for the age, profound in its complexity and very scientifically poignant. If you look closely at *The Keys*, you will even see very specific magnetic information about DNA!

The Seventy-Two Names of God is also a very popular subject these days, especially within the study of the Kabbalah. If you go to the Internet and look for a book by that name, a very popular one will come up from another author. Hurtac's was different, and the names are different. Let me quote from the introduction in Hurtac's book:

> "Be mindful that the Enochian use of the Sacred Language is based on a slightly different tonal synthesis of the ANCIENT Hebrew in conjunction with the Aramaic, the Egyptian, the Palio-Semitic, and Phoenician nuances of the pre-New Testament period."

Hurtac was told that the fuller meanings of the Sacred Expressions cannot be revealed in the wisdom of man, and that whomever asks, an explanation will be given according to their state of consciousness. All of this falls into perfect sense as I capture 12 of the names and have an angelic entity give me the definitions, just as Hurtac said it should be. I personally find it sad that his profound book is no longer in print. I think it's the best of any I have seen, and a beautiful book at that.

The Hebrew Issue

There I sat with all this Hebrew. The first thing I wanted to know was why Hebrew? Isn't that a fair question? Kryon is about to reveal the 12 energies of DNA and they have Hebrew names. Why not Lemurian or Sumerian? Hebrew is not the oldest language we have, and I was more than curious why (I also didn't want to learn any Hebrew). From Kryon, the answer came that Hebrew is the core language of a monotheistic God. Abraham is the acknowledged father of the three major religions on the planet. He spoke ancient Hebrew, so this is what Kryon meant. It keeps the teaching within a context of a still current Human language and honors the beginning of a "one God" experience for humanity.

That's all well and good, but I wasn't prepared for the onslaught of opinions about Hebrew in general! Oy vey! (I know, I know, that's Yiddish.) It seems that everyone who speaks Hebrew is an expert in Hebrew! What to do? Who should I turn to for a good, generic interpretation of the words and the pronunciations? The answer was right in front of me and one that plays very significantly into this book.

Enter Elan Dubro-Cohen

I've been to Israel a number of times now and channelled Kryon for some very large groups both in Tel Aviv and Jerusalem. I remember the first time the Kryon team went there in the year 2000 during the uprising of violence. Tourism had about dried up and the customs officials pumped our hands in appreciation of being some of the only American tourists who had braved the bad publicity (and occasional bombings) to be there that day. I felt that if the ordinary citizens of Israel could do it, then I could do it.

I was met by my host, Elan Dubro-Cohen. He was born in Israel, had been there all his life, and was the perfect host. This is typical that we have hosts of Kryon seminars who are local to the area. As a graphic artist, Elan had also designed most of the Kryon book covers for the Hebrew language. Therefore, he was not only my host, but also a wonderful artist who was very familiar with the Kryon energy.

One of our permanent Kryon team members is Peggy Phoenix Dubro. She provides profound healing/balancing work on our planet, and it began about the same time as mine did, back in 1989. Peggy is a seeker and by 1989 had been through many of the protocols of spirituality available on Earth when she began receiving messages about what would become the EMF Balancing Technique and another profound teaching called Lattice Logic. (Yep, it was channelled to her.) These profound processes, now with many advanced levels, are current in more than 50 countries worldwide and continue to expand. I was fortunate to have her joining me in Israel and her family's presence there would be, well, uh, very synchronistic. Elan fell in love with her daughter! (Queue the romantic music here.)

Ten years later now finds Elan living in Sedona, Arizona, with his beautiful wife, Shana, and their daughter, Ahdivah. He is now also a naturalized American citizen! Elan is a vital part of the EMF worldwide work and is also the ongoing host of the American Kryon flagship meeting: The Kryon Summer Light Conference in Sedona, Arizona, given in June every year.

It is Elan who I turned to for Hebrew help, and the one who is also the designer of the cover of this book. So as far as Hebrew is concerned, I was getting a native! Anyone could argue Hebrew spelling now, but they would have to do it with a fellow Israeli. Hebrew is his native language and he had spoken it all his life. So Elan, a native born Israeli and Hebrew speaker, was just an Arizona phone call away! I began my lessons.

Elan jumped right into this whole subject. In fact, he got some major "downloads" of his own when he began to see the names that Kryon had given me. Even before we had all 12 channelled, Elan was designing beautiful paintings that represented the energy of each layer. It was perfect (and still gives me the chills), for here is a Hebrew speaker getting the whole picture, even before I do, of the profundity of the Hebrew names and what they represent to him and his culture.

Within the pages of this book is the channelled artwork for each of the layers of DNA by Elan. If you like them, they are even available as a beautiful poster set. More information about that is in the color section of this book in the back pages. I never expected this book to have such an artistic flair, but then life is filled with surprises for me, and this was one.

What the Illustrations Are About (pages 321-332)

Since this book has been in progress for a number of years waiting for the energy of the release (obviously waiting for 2010), the DNA illustrations have been available for some time. Many have asked, *"What should we do with them? Do they power something? Do they 'do' something?"*

The answer: Think of them as a 3D representation of the mastery within you. Look upon them as a reflection of what you carry inside. Look at them and smile. Do they have power? Let me ask you this: Does a photo of your child have power? The answer is that it has the power you give it, based on your own situation, and your own illumination of what it is that is meaningful to you.

It's just a beautiful reminder of who you are and the teaching in this book. Don't worship these illustrations, or create an alter with them, or start to carry them around with you hoping they will do something for you. Funny enough, there are those who are doing that already! The truth is that if you "feel" something around them, then you are receiving a wonderful message of remembrance from them. Smile and just appreciate these works of art.

Regarding Hebrew, I found out that it's hard! I have a Southern California language culture and so that hair ball sound is especially difficult for me! OK, OK, if you are a Hebrew speaker, calm down. That was a very inappropriate remark, but I'm known for being inappropriate, so it stands. Let's laugh when we can at our own weaknesses—and mine were obvious. I was not able to appreciate this fine, ancient, and complex language that had sounds I liked to make fun of. Don't send me letters. I use them for paper airplanes that go directly into my furnace.

As each of the DNA layers were given by Kryon, we placed them on the Web site. They are still there, and you can see and hear them at [www.kryon.com/DNA].

We also recorded Elan saying each name and placed it on the Web, so you can hear them as well spoken by an Israeli and also see a pronunciation guide for each one. In the process of learning some of these works, however, Elan began to inform me that we would get letters about the names, what they meant, and how to pronounce them. Boy, was he right! It seems like Hebrew speakers are very particular and opinionated about this core language. There would be many opinions. Perhaps you have one of your own? Perhaps you don't agree with the pronunciation or how we have treated them? Well, if that's the case, you can e-mail Elan (smile). Speak Hebrew to him, too (more smiles). His e-mail is in the back of the book.

For Hebrew readers

Kryon choose the Hebrew names in regards to their close resonance to the layer vibration/energy itself and not the literal meaning in Hebrew, although sometimes you can find similarity with that as well. The reason Kryon choose Hebrew names in the first place to represent the DNA layers, was mainly to honor the Core Energy of the Earth.

My test came in June 2007 when I presented all 12 layers of DNA at our Summer Light Conference in Mt. Shasta, California. I had to pronounce each name, followed by Elan doing the same (gasp). So during that meeting, the attendees had an immediate comparison of what I had learned against the Israeli-born guy saying it. My friends tell me that it went well—but again, they are my friends and they don't speak Hebrew. The Hebrew speakers in the audience were also kind, calling my attempt "interesting" and other words that really meant, "You shouldn't be doing this."

SUMMARY

So as we begin the study in the chapters that list the layers, you now know where the words came from, why they are in Hebrew, and where to hear them correctly. You also know that the actual

Hebrew translated meanings are not necessarily the Kryon meanings, and almost never are they the Hurtac meanings. Some of them even have two meanings! (Getting complex here, I know!) This is all appropriate, since the subject is DNA, not Hebrew. Speaking of that, let's move on to DNA itself.

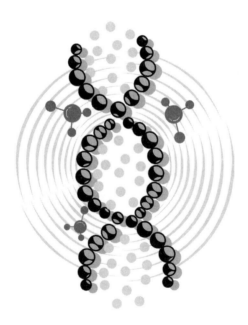

Chapter Three

DNA Examined in a New Way
Kryon
Book 12

Lee Carroll

Chapter Three
DNA Examined in a New Way
by Lee Carroll

This is not going to be very scientific. In fact, this short discussion is really for people like me who never took chemistry, did poorly in biology class, and need a "dummies book" to get through some of the new math... okay, the old math too.

DNA is a huge mystery to most of us. We hear the name, kind of understand what it's about, but really have no idea of the complexity or how it works. So let's look at some of the "gee whiz" stuff that you might not know—first, the things that are taught in school, then the things that are not.

DNA (deoxyribonucleic acid) is the core blueprint of a Human Being. The building blocks of DNA are called DNA bases and there are four different kinds of these. It is the sequence of DNA bases that makes up the genetic code. In the famous double helix shape, the DNA bases pair and twist around each other many times in a spiral form. The Human genetic code is about 3 billion base pairs long and occurs about a hundred trillion times in a typical Human body. The Human genetic code contains approximately 23,000 Human genes (at last count), which represent the propensities of who we are. This includes everything we get from nature, plus what we get from our parents and their parents. A gene is the heredity unit in a living organism. It may include eye color, as well as how to build a kidney, as well as the propensity for health weaknesses down the line, such as Alzheimer's disease. So "heredity" is not just the attributes from your parents, but the general attributes of a Human Being as well. You might say that the genes carry forward all that we are that is Human, with unique specialties that are from our parents. You may have heard of RNA (ribonucleic acid)? It's part of the process of transcription, which is the beginning of the actual production line of protein creation and orchestration. But the DNA is the instruction set, and that's what we wish to speak more about.

So, follow me here: This very tiny double helix molecule is only visible through high-powered electron microscopes. The whole double helix is a chemical system that represents 3 billion parts! Can you imagine how small those parts must be? We only really saw the actual shape of DNA as deduced by Watson and Crick (an American and a UK combo) in England in 1953. I wish to interrupt this discussion by also giving Rosalind Franklin kudos. Who is that? This is a woman who was a specialist in X-ray diffraction at Cambridge University. In all fairness, she was also working on the same issues as Watson and Crick. Watson was inappropriately and secretly shown a very "telling" X-ray picture from her work, and he realized the shape of DNA instantly. Without mentioning her contribution, the two men got all the credit for the discovery. Her name seldom occurs in conjunction with DNA discovery, and she died in 1958 of cancer. So there are not many champions of her involvement. Let me be one. Now you know.

It took another 43 years until February of 2001 for science to "map" the whole DNA molecule. It's a fun story that should be made into a movie. The government created a project called "The Human Genome Project" and began mapping DNA chemistry in 1990. They slowly processed things in ways that only government can (smile). In the meantime, biologist-entrepreneur J. Craig Venter had a better idea and started his project in 1998 using a very unorthodox method called Whole-Genome Shotgunning. Both projects announced the results together, meaning that Venter beat the government method by almost eight years!

Then the real work began, for the mapping only revealed the basic chemistry of DNA. Think of it as "letters" of a giant book. Now they knew each letter, but they had no idea what the language was! They needed the language to give them the bigger picture, which would give them clues as to the words in the book and where the genes were. This is when they discovered they were in for a ride they didn't expect. This is also where we begin to draw comparisons to what we are teaching in this book. For the best science and computers in the country were hard pressed to find the codes they expected within the chemical structure of the Human genome.

We think in 3D. We can't help it. It's our reality and we can't be expected to get past it. But it often stops us from seeing a bigger picture. Science is now beginning to shout to us that our Universe and all that is in it is multidimensional. So eventually we are going to have to create math to match that, as well as new physical laws, as well as an expanded thinking. But for now, the scientists were assuming a very big thing—that the Human genome was linear and that the whole gene structure of a Human Being would be spread across the 3 billion "letters" of DNA. It wasn't. The reason I'm including four channellings in this book is because they are so germane to this entire discussion, and one of them is Kryon's discussion on the 3D bias of science. Although not specific to DNA, I think you will enjoy it. (page 280)

Back to the subject: Against all logic, the scientists couldn't find the codes they absolutely *knew* were there. They used some of the best code-breaking computers of the day, looking for symmetry of the kind that any languages would create. Then they found it. It must have blown them away, and at the same time gave them one of the largest biological puzzles of the century.

Within all the letters, all the chemistry, and all the obvious complexity of the Human genome, only about 4 percent had a code!

Only the *protein encoded* DNA chemistry carried a clear code for the making of genes, and when it was there, it was really obvious. It was so 3D that you could see "start" and "stop" markers for gene sequences! Much like computer codes of the day, the chemistry was mimicking what we expected, but only a small fraction of the Human genome was involved in making all 23,000 genes of the Human body. The rest seemed to do nothing at all.

Let me give you an example of the frustration: A flying saucer appears overhead. It can do amazing things—hover, void gravity, and all those good things we expect from a flying saucer. It lands and sits there. We approach it and realize that nobody is inside. It must be a robot, having landed itself. Suddenly, the "hood" raises and invites our best scientists to examine how it works. We are very excited, realizing that there must be secrets ready to be discovered. New physics is about to be revealed! Then we start examining the engine and are shocked; the engine compartment is stuffed with a complex array of packing material. It's interconnected, and some of it even moves, but it does nothing. There is no "system" at all, and it's like random packing "popcorn" that you get when something is shipped to you. You dig and dig, tossing the popcorn out, and suddenly a tiny item appears that is shiny, connected with wires seemingly to nothing, and obviously is the core of the engine. You can hold it in the palm of your hand, and it controls everything! Then you start experimenting. Oops! When you remove the packing material, the saucer doesn't fly. When you put the packing material back in, it does! But the random packing material doesn't really "do anything"—or does it? How can packing material do anything? The bias is clear: We expected an engine—something shiny, connected by wires, linear, complete with structure—and we found it. What we "saw" as padding or packing material, we immediately threw out. Do you see the bias here, and the metaphor?

It became a joke. DNA was composed of 3 billion parts, most of which did nothing! Only a tiny 4 percent did it all! That didn't make sense! We know that nature is very efficient. We can see the evolution of living things even within our own lifetime, and we realize how efficient nature is. If fish are trapped in an underground cave, within 10 years or so, you will find that they have lost their eyes. Nature throws away things it does not need, and we can see this everywhere. Yet we seem to have DNA with 96 percent "junk"! We are supposedly the top of the evolutionary chain—and we are 96 percent junk? It was against everything we saw in nature, yet it was there. Therefore, the non-protein encoded portions of DNA were labeled "junk" even by some of our most prestigious scientific minds. The non-protein encoded parts were random, no symmetry, no purpose, no use at all.

Enter the Non-3D Thinkers

You approach the flying saucer with a new idea. Perhaps the "packing material," which seems to be random, isn't an engine at all! Perhaps it's a map! After all, the craft has to know where it's going. Then you think, perhaps it's a different kind of map. Could it be that in a quantum state, the craft needs to have a quantum map? What would that be? It would be something that helps it to exist in linearity, yet provides instruction sets to the tiny, shiny part of the engine to allow it to control the craft in 3D. In this example story, we know the craft has multidimensional attributes, for it can control its mass. We also know from our own quantum physics that when we move into a multidimensional realm, time and distance cease to exist as we perceive it. These things are instead replaced by potentials and a full nonlinear and confusing array of "event rules" that make very little sense to us in 3D. Therefore, the odd "random" packing material isn't random at all; it just looks that way to a 3D creature! (you and I and the scientists) It's needed exactly where it is to "posture" the tiny part of the engine to move the craft. You

Chapter Three

might say that packing material is the "modifier" of the engine, and there has to be a large portion of it, since there is so much to "tell" the engine how to move in a multidimensional way.

For years, "junk DNA" has been an expression that we have lived with. Suddenly, however, we began to think differently. *"What if...,"* someone postulated, *"there is no code in the junk because there doesn't have to be. What if the entire 96 percent of DNA somehow were nonlinear quantum-like rules that 'steered' the coded parts?"* It's a brand new controversial concept—but at least it's a concept that is beginning to think out of the box of a constricted 3D logic!

Here is a July 13, 2007, report from UCSD, reported by CBS News:

"So-called junk DNA—the 96 percent of the Human Genome that seems to have no useful purpose—may play a more valuable role than its name suggests, U.S. scientists say. Now an international collaboration of scientists has found that some of the 'junk' DNA may serve to create boundaries that help properly organize the other 4 percent. 'Some of the junk DNA might be considered punctuation marks, commas and periods that help make sense of the coded portions of the genome,' said co-author Victoria Lunyak, an assistant research scientist at UCSD."

My opinion is that we are seeing a multidimensional aspect of our biology that is obviously huge! What if 96 percent of our DNA was a seemingly random instruction set for the 4 percent? It's not random at all, but just appears that way to a 3D mind-set. Would punctuation seem to be a letter in the alphabet? No. Then what is it? Does it have symmetry? Does it spell anything? No. If you look at punctuation marks in our language, it actually seems to be random. If you looked at this page, for instance, and didn't know the language or how it worked, the punctuation marks would seem to make no sense at all. They have no symmetry. If you put this page through a super computer, it would recognize the words, eventually, and what they probably said, but not the punctuation marks.

Think about it. The engine we looked for in our saucer was, indeed, there. The 4 percent protein-encoded parts are the "shiny engine". The "junk" is the 96 percent that looks like packing material. Now we suspect something entirely different is taking place, where the 96 percent might actually be a multidimensional designer's template, and the 4 percent is just the engine that obeys the design.

Isn't the ratio interesting? When we begin the esoteric teaching of Kryon later in this book, you will immediately see that in Kryon's teaching only 8 percent of DNA is in 3D (layer #1). The other 11 are multidimensional. In Kryon's scenario, 92 percent of DNA modifies all the rest. By the way, in Kryon's scenario, everything is base-12 (smile).

We may actually be seeing a slow recognition that DNA functions far differently than we expected, and with much more complexity than just a code one can read in chemistry. When you see what Kryon has to say about the other 11 layers, you begin to see what might actually be in the "junk." This is when you will laugh at just how 3D we all are.

Chapter Four

Things That Nobody Thinks About
Kryon
Book 12

Lee Carroll

Chapter Four
Things That Nobody Thinks About
by Lee Carroll

DNA Communication: There is much more information about DNA communication coming to light in recent discoveries, and some of it plays into other areas I teach. One item is that DNA has been discovered to be a loop instead of a strand. This allows the idea that DNA might have a current running through it and actually have a magnetic field of its own. In 2008, I was told by a scientist that his lab had proven just that! So suddenly we have a picture of DNA with its own magnetic field, if he was correct.

This same scientist made a statement to me that I'll never forget: *"Watson and Crick did a huge disservice to mankind when they showed the double helix the way they did."* I was taken back. What does that mean? He went on to explain that everyone thinks that within Human biology there are trillions of little "ladders" floating around in our molecular structure—ones that are straight, just like the illustration of DNA that Watson and Crick showed us.

The truth is that the double helix is a schematic of chemistry, not a picture of DNA. *"You mean it isn't a ladder, or a double helix?"* No, I don't mean that. What this researcher told me now completes a puzzle of metaphysics and mainstream science that very few think about; he told me that his research showed that DNA clumps around each other in complex ways. It's the double helix, but it's also ladders wrapped around ladders wrapped around ladders.

In electricity, there is a phenomenon known as "inductance." This is when one magnetic field overlaps another. In this mixed field, we get the "magic" of "special communication" and a whole bunch of other electronic attributes that we use every day. Inductance allows for the amplification of voltage due to turn ratios of the cores of transformers. It allows for communication of signals through magnetic fields without wires and without any added noise or anomalies. And one of the most common and profound uses of inductance is the creation of small transformers (creating magnetic fields) that allow for signals to be sent along miles of wire without the addition of noise and without loss. It's magnetics at its best, and we use it all the time, especially in the communications industry.

So then, if DNA has a magnetic field and its natural state is to be intermingled in a nonlinear way with other DNA, it would begin to explain something that we almost never think about: How can trillions of molecules of DNA communicate so that they present one consciousness of a unique Human body? Did you ever hear that question in school? The answer may just be "inductance."

We learned in school that it's "just part of the system of the Human body," but the actual communication isn't discussed. But with the possibility of magnetic DNA, and also the idea that 96 percent may be multidimensional "instruction sets," it's quite possible that it's the magnetic attributes of DNA that allow master communication at a level that is mind boggling. After all, every molecule knows the "plan" of you and they all "talk to each other"! They have to.

Birth- The Hidden Question That Few Ask

Two children are produced by the same set of parents. One is a master musician and the other can't carry a tune. One is self-assured and the other is afraid of his shadow. They each have different phobias and fears, but they have the same eye color and skin coloring as the parents. This scenario is repeated millions of times a year and nobody ever asks, *"What gives here?"*

Science tells us that when we are born, we inherit certain attributes from our parents. I believe that. But what about the "other stuff"? Where did the talent come from that a parent never had or the fear? Why is it that one child can figure out how almost anything works, and the other one wants to be a painter? Birth after birth, we see a system that literally shouts that there is something more than what our parents biological contribution to who we are.

Psychologists tell us that the brain is filled with amazing variety, and that this drives the kind of potentials of Humans at birth. Kryon tells us that there is a system: When birth occurs, we have the following things imbued into our DNA:

(1) The heredity of our parents' biology. This is not only eye and hair color, sizes and face attributes, but also some talent for certain things (or not). This is the 3D chemistry in the double helix and represents the parts that carry over attributes from our parents and their parents.

(2) Karmic imprint. This is what many ancient religions have taught: unfinished business carries over from the immediate past life. This can be positive or negative. For instance, have you heard of the child who paints like a master or plays the violin at three? Where did that come from? It's a continuation of the last life. It also can carry with it an individual who is mad at the world and depressed all the time. He/she is continuing a process that he may have been working on for several lifetimes—that of finding self-worth and peace within. Do I have to mention guilt? It seems to be an attribute of certain cultures in all their children. This is karmic, not chemical. That means it's *information*, not chemistry.

(3) Astrological imprint. This is obvious. It's a snapshot of the pattern of the solar system at birth. Astrology is the oldest science on the planet and is probably very related to gravitational and magnetic imprints, which are a multidimensional patterning of the

Chapter Four 53

sun. The incredible push and pull of the planets creates this pattern at the sun's core. This pattern gets transmitted to the earth every moment through inductance with the solar wind (heliosphere). You can see the place where inductance actually happens through the observation of the aurora borealis, where the heliosphere overlaps the earth's magnetic field (you actually get sparks!). This places the sun's quantum imprint onto the magnetic grid of the planet. The magnetic grid, which overlaps the magnetics of your DNA, then patterns a portion of your DNA at birth (again through inductance). If you didn't follow this discussion, don't worry. It's not important to the book. But for those who did, it's the only place you will ever have seen the potential scientific explanation of astrology in a book.

Many don't believe in astrology, but are the first to acknowledge the Human crazies of a full moon experience. Police put on extra shifts and hospitals adjust. It's the force of planetary objects and their effect on Humans. Nobody thinks much about that. It's easier to just deny it and work through it than to actually connect the dots and admit that there is something going on. It might be explainable through a further understanding of multidimensional forces. My prediction is that this quantumness will occur someday and be acknowledged. It has to be, and if you doubt this, just look at what they have spent on the large Hadron Accelerator in Geneva. All of this represents the search for multidimensional things and the acknowledgement of new forces outside of Einsteinian and Euclidean physics... invisible matter!

(4) The entire Akashic Record of all your lifetimes on this planet. This is probably the most profound of them all, since it carries a residual of who you were, and it works with the karmic portions of energy mentioned above. Kryon continues to teach that the attributes of these "past" lives are available today, and that's why they are there at all. He encourages us to "mine" the Akash—that is, to discover what we have within our own DNA's history and use it.

(5) The sacredness of God as represented by your Higher-Self—an energy given to humanity by the Pleiadians approximately 100,000 years ago. This is the subject of the next section of this chapter, and is the most esoteric and controversial thing you will read in this book.

Now, take a look at what you just read. Only item number one is linear and observable under the microscope. Everything else is the "map" for you and I, represented by multidimensional energies, the existence of which science is now starting to acknowledge.

The Pleiadian Connection: Truly Weird!

In metaphysics, there are often very startling and unbelievable premises. This is especially true if you are new to it. Those who have been in it a long time don't flinch. But some of the attributes of how humanity got started are hard to believe, since there really is no science to back it up.

On page 258 of this book I present a channelling entitled, "The History of DNA and the Human Race." If you make it that far, then all the information within that chapter fourteen will be already revealed. But I wanted you to hear it from Kryon directly, as well as in this personal article, so you'll have a better idea of what Kryon has told us.

What I discuss next has little or no scientific basis, but more than a billion people on the planet are just fine with the allegory of Adam and Eve. This, or course, has even less science behind it but it's just fine since it is the core mythology that we find in the Bible (written by Humans, remember?). It's interesting that we will fight wars over this mythology and "take it to the bank," as they say. But let someone like me, for instance, talk about something else that might have happened and those in the various religious boxes find it totally unbelievable. They will pound the table and yell and say,

Chapter Four

"*What's wrong with you? The Bible clearly tells us what happened!*" So I ponder the statement without a fight. (My belief system is not evangelical and invites a place for all spiritual thinking.) But my inside just wants so bad to say, "*OK, you can keep your talking snake story. I'm sorry, I don't know what came over me.*"

Evolutionary information changes all the time. Once we thought that the Grand Canyon took millions of years to create. Now, we are not certain. It could have been only a few hundred! Geology itself is changing greatly, and with it many of the anthropological beliefs are being challenged as well. So let's just say it—here is something that nobody ever thinks about:

There is Only One Kind of Human on The Planet

"*What?*" some will say. "*There are many kinds!*" Not really. There are many colors and sizes, but only one kind. Gone to the zoo lately and looked at the other mammals? What about those who even look like us? There are seemingly dozens of kinds of monkeys—sizes, hair color, shape, with and without tails, etc. If you look at almost all the other animals on the planet, there are many "types," but Humans have only one type—the one you are looking at in the mirror.

Nobody thinks about this, since it's so common that it's out of sight and out of mind. But it truly is counterintuitive to the whole method of Earth evolution. Think about it. How is it that we only have one Human type when evolution naturally creates many types? The animals around us show it, yet we don't fit this evolutionary mold at all. How could this be?

Kryon has given the answer many times, and started in about 1990 or so telling us about the "seeding of Human DNA" with spiritual attributes from afar. It's the story of the Pleiadian connection and the beginning of Lemuria. Kryon's statement is this: "*Don't go back in history any further than 100,000 years to find Humans who are like you."* Anthropologists tell us that Humans were in development long before that, so what's Kryon talking about? It's about an extremely controversial premise that has our Human DNA being changed by those from another star system about 100,000 years ago. (Please don't put down the book yet. Wait at least until the end of this section.)

Kryon has told us that beings from another world, who look much like we do, came to this planet over a period of many years and literally interfered (appropriately) with normal Human evolution. They gave us the layers of DNA that we now see as spiritual. They enabled our entire race to be part of the spiritual plan of the Universe, and allowed us to become "pieces of God". If you tell this to other people who are not metaphysical, they will run the other way. The Adam and Eve story is somehow far more acceptable! The apple, the snake, God's garden, the devil, Human innocence is all a much more believable story than something to do with ETs. (It seems that a devil story always wins over an ET story.)

This Pleiadian story does not belong to Kryon. In fact, it has been with us for a very long time in other cultures, and even perhaps from the beginning. But in metaphysics, it has been told and retold by authors and channellers since I was born. Along with it, of course, comes those who wish to take this beautiful story straight into drama—wars between worlds, controversial "takeover Earth" scenarios and a group of lizard folks ("Lizzies") that hide beneath our skin. I know this because I had the honored distinction of actually being called a "Lizzie" in a public forum. My, that was such a proud day in my life (sigh). This, by the way, is so common with almost everything esoteric that I hesitate to even comment on it. It's what Human nature does, and many just love to be scared and spin around in the drama of it all. Conspiratorial energies are far, far easier to manifest than ones where we might actually have to look within ourselves and find truth. They are the ultimate distraction from maturing. My "Lizzie branding" experience was actually quite mild in comparison to the three-month campaign by a woman in Tucson who labeled me the "Evil of the Century" about 15 years ago. So it was mild and kind of cute.

Let's Look at Some Synchronicity

I sat in a circle with the Hawaiian elder "Grandfather" Hale Makua in 1990 on the island of Kauai in the Hawaiian Islands. The Hawaiians, much like the American Indians, pass down their history verbally through the young men of the tribe. They tell them the same stories over and over from the time they are able to comprehend, and they depend on this method to make certain their history is known correctly and will continue to be passed on accurately. There are basically two reasons for not putting this in some kind of written archive: (1) Not all the tribes were literate when this method began, and (2) To keep it away from those who are not authorized to hear this sacred information (those they don't trust, like us).

Makua was a modern man, having served as a Marine in Vietnam. But he was also destined to be a spiritual leader for his Hawaiian people, and many turned to him for wisdom. Upholding the tradition, he taught the young Hawaiian men about their Polynesian history. The focus, by the way, is always on the ancestors. My presence in an outer circle of non-Hawaiians was a gift from Makua, and a statement that he really wanted others of like consciousness to begin to know what he knew. The pure Hawaiians are facing a challenge that many of the indigenous are also seeing; the modern world is beginning to cull out many of the tribe. The pureness of the race is being cut by intermarriage, and modern culture is carrying away those young tribal men to jobs that pay more, instead of them staying in the tribal areas. Almost worldwide we are seeing many of the indigenous grandfathers begin to trust more and more of us who are not indigenous with some of the sacred secrets of their history. This isn't an easy decision for them, but it may very well save the record of their people, what they believe, and how they lived.

The balmy afternoon had turned to a warm early evening as a number of us in the invited outer circle continued to listen to Makua. There was a great weight placed on the ancestors of the Hawaiian people. It was Makua's way and the core of his teaching, for he wanted all to understand that the ancestors had the pure knowledge and they also had the wisdom. It was up to the young men to capture it and pass it on. Then Makua said something that jolted me: *"The bodies of our Lemurian ancestors are buried nearby."* It's the first time I had ever heard the word "Lemuria" used by a Hawaiian teacher! Lemuria was supposed to be the first large civilization of man, according to Kryon and other spiritual historians. It lasted for many thousands of years, mainly due to their isolation. Lemuria was located around, and on, the largest mountain on Earth—the largest not in altitude, but as measured from the bottom to the top. That's Hawaii. It's one mountain, currently situated on the bottom of the Pacific Ocean, which has various peaks protruding from the water and are known as the Hawaiian Islands. Kryon told us this, but I didn't know this was also Polynesian lore.

I got very interested in what else Makua might share that could meld indigenous history with our own metaphysics. So when it was appropriate, I asked Makua a burning question: *"Master, what is the Polynesian 'Adam and Eve' story? In other words, how did humanity begin?"* Makua's answer knocked me over, and thinking about it still gives me chills to this day. He stood up and slowly looked at the sky. When he found what he knew would be there, he raised his arm and his hand pointed to the seven sisters constellation. *"The canoes came from there!"* he said. Makua was pointing at the Pleiades.

Chapter Four

Could it be that ancient Hawaiian history, as taught from the earliest of times, had the same unbelievable ET story as the one Kryon told me? It shook me to the core, for here was my personal proof that what Kryon had said about the Pleiadians was true. They were the masters of our spiritual DNA shift, and this means we will probably meet them someday.

My good friend, Woody Vaspra, and his partner, Catie, are very involved in the task of creating an indigenous wisdom council for the planet. He is president and elder liaison in the World Council of Elders, and he and Catie co-founded this organization. It's slow going, but in the process, Woody, a pure Hawaiian, has made the journey of years it takes to become recognized as an elder. In Hawaii, they call him Kupuna (elder), which is a title he humbly does not use himself when he is there.

Due to his indigenous work, Woody has been recognized as an elder in many other native circles. The most notable is in his efforts with the Oglala of the Lakota Nation. It was through his process of adoption and participation in the *Sundance* that they offered Woody the position as Leader of the International Sundance. You must be an elder for this. In my discussions with him, I now realize the profundity of the Seven Sisters Hawaiian story, for Woody informed me that the Hawaiians even have a name for the Pleiadians: Makali'i. There is even a sacred day for ceremony in the months of October and November when the Seven Sisters is directly over Hawaii. All this information I have received even after the death of Makua. I thought I had lost my elder Hawaiian connection, but Woody has rekindled it!

In November of 1998, Kryon channelled that from about 100,000 years ago to about 50,000 years ago, the many kinds of Humans began to slowly diminish. One kind had been given the DNA change by the Pleiadians, and this one kind had slowly replaced all the other kinds. It might be due to intelligence or increased comprehension. It might have been that the "one kind" was better at survival. I don't know this, and it has never been revealed to me. But this information about there being only one kind of Human remained as just something that was channeled—until December of 1999 when I actually saw the same exact scenario on the cover of *Scientific American*! (Validation music goes here!)

I was in an airport when I saw the cover. I almost fell over! There on the cover of *Scientific American* was a drawing of several kinds of Human Beings. The title of the article inside was between them: "Our species had at least 15 cousins. Only we remain. Why?" (*Scientific American*, January 2000). I have presented this before, and you can actually find a copy of the cover of this magazine in Kryon Book Eight on page 369.

Here it was—what nobody ever thinks about, now openly being questioned. The article inside told the same story Kryon gave us. It seems that we had a good start, like all the other animals, at having a great variety of our kind. There were ones with tails and without tails, etc. But about 100,000 years ago, something happened (according to the magazine) and all the others began to die off, leaving only the kind we have today—you and I.

This is a "confluence of synchronicity". In other words, many things had come together in a mainstream 3D way that now showed me that the Kryon story was more accurate than ever. I don't really expect mainstream science to ever agree with the Pleiadian story until the day we actually are visited and a very tall Human figure gets out of a ship and tells the same story. Even then, many won't believe it.

The synchronicity continues, for what is next was written for this book almost last due to something that happened to me—kind of a punctuation mark on the whole Pleiadian subject. I was stunned

Chapter Four

when I made a discovery, one that is probably well known to many, but had to be presented to me in a certain way as proof, in a sacred place, in the middle of the Australian continent.

I love to visit the special places of Earth when I can. In the early days of Kryon, I went to Australia three times, but I never really saw much more than a few sites in almost every one of their fine cities. Even in New Zealand, I didn't really appreciate what was there—one of the most beautiful scenic wonders of the planet. It was only later in my journeys that I decided to stay and see some of the incredible places of Earth, many times alone, other times with some of the Kryon team. I'm not speaking of the cities I present in, but rather extensive side trips to unusual places that I might be close to while presenting in the larger metropolitan areas.

When planning a trip back to Australia for the fifth time in March 2010, I wanted to finally see the wonder of Ayers Rock. This is a huge, sacred sandstone monolithic, a bright orange rock sticking out of the earth right in the middle of the Australian continent (part of the northern territory in central Australia). Around this mountainous shape are the same red rocks and red dirt that you find in Sedona, Arizona, and the 1100-foot high monolith is one piece of elegantly grooved sandstone, making it a geological oddity. Due to the incredible orange color of it, it is also a photographer's dream, and well documented and presented in photos all over the world. It's also an energetic center, kind of the navel of Australia, in a place that has just about zero resources of its own. Everything has to be brought in from the coastal areas—everything.

Other attributes were that this is one of the hottest places in the Australian outback (often up to 118F), filled with giant bugs (to me they were giant), and more flies than I had ever seen in one place. Ants were everywhere, some of them very large, and so you have to watch where you stop since they often quickly try to take up residence in your pant leg. Termite mounds abounded and we were told that there were more wild, single-humped camels in that area than anywhere in the world (didn't see one). However, despite all these obvious outback attributes, it is a very popular place with good accommodations and strong air conditioning. Paths are clearly marked, roads are good, free transportation is provided, and there is a lot of good help for the tourists. A full resort was there, swimming pools and all, and I just had to go. So I got my tickets ready as soon as they told me I could have a fly net for my hat (a stylish outback hat, of course). The photos are on my Web site if you wish to find them.

Better known these days as the town of Uluru, the government had given the whole area back to the indigenous people in the late '80s, and a kind of "give and take" was present between the Aborigines and those running the tourist industry. The sacredness of the rock was being honored in some places to the fullest, yet being violated in others. Uluru has its own airport with one runway, surrounded by that beautiful red dirt. Flights were now possible direct from Sydney, something only just available since 2009.

Many places around the rock are so sacred to the elders that they are closed to tourists, and photos are not possible in certain areas. Helicopter rides are only allowed for full rock viewing during early morning hours (for light reasons), and then are restricted in the afternoon (kind of a compromise to the sacredness of the rock versus those who wish to photograph it). There are even rules about female helicopter pilots not being allowed to fly over certain areas (again, following the wishes of the elders of the indigenous tribes who are in charge of the area and who are sensitive about the roles of men and women within their culture).

However, in all this, much to the chagrin of the elders, tourists are still allowed to climb this sacred rock! The elders are in charge,

Chapter Four

and continually close the path during the day for their own reasons (probably to discourage planning a climb), but the climbers persist. Perhaps in time that will change, and those who enjoy climbing can scale something else close to it instead (there are other similar rocks within a few miles of it).

So, it is March 2010 and we are an eclectic group of "Kryonites" who are there to brave the flies and ants and hot sun. I was travelling with a few Australian Kryon fans (one of them an experienced park ranger) and Jorge Bianchi (my South American coordinator). After taking off from Sydney, we landed about three hours later to spectacular weather! It was not hot, and had actually been raining! (Seldom does this happen there.) So, against all odds, it was pleasant and green (thank you, God)! However, the flies were still there. (They have their own convention every day, it seems.)

I remember walking into the hotel registration area with our troupe. A giant centipede was crawling undisturbed in the middle of the lounge, on his way into the kitchen area (probably to get a cup of coffee). Nobody seemed to care, and it was the first indication that bugs are king and we are just visiting their domain. Dozens of moths and grasshoppers were lounging around, sitting on the guest chairs and tables overlooking the pool (sigh). If you displace them and use the chair yourself, eventually they will be on your lap. I guess there are many of these kinds of places on Earth, but I try to stay away from them. The second star in a hotel rating is supposed to represent a bug-free environment... somebody told me that. I don't know what the first star is—probably that they actually have a building.

Our first and most profound excursion was a trip to the indigenous area, their museums and shops, and also a tour given by a senior elder of the Anangu tribe, responsible for the cultural center. The elder's name was Sammy. Sammy understood and spoke English (kind of), but he used a young Aborigine interpreter and wished to give the tour around a certain portion of the rock in his native tongue. He was filled with family stories of his long lineage in the area, some even from the 1800s. Many of the tales would take your breath away—some joyful, some sad.

Funny thing about the flies. Did I mention there were flies? They were relentless. If you didn't have a net, they knew it and no matter how many times you swatted them, they would come back with their relatives for more fun. It was "irritate a channeller" week and they were on to me! But Elder Sammy didn't have any around him! I watched, and they simply left him alone. Go figure! I also saw them ignore our Aussie ranger, too! Maybe they just know that if you are a tourist from overseas, it's a lot more fun for a fly. Screaming, swearing, giant hand motions and running around to escape them must actually attract them. With all of them swarming around, I did manage to eat one, but unfortunately it was at an outside nighttime formal dinner affair—white tablecloths, gourmet food, people dressed up (sigh). As I was drinking my water (with the gourmet fly in it), I immediately spit it out onto the tablecloth with great "get that live, wiggling giant fly/bug out of my mouth" gusto. Sometimes even celebrity channellers do inappropriate things (double sigh). It keeps me in my place. I immediately stood up, tripped over my chair, and fell in the dirt (very red dirt with very white pants on). Many in my group felt this was the highlight of the evening (triple sigh).

The next day, I was wandering around the cultural center by myself with my fly net in place, reading the history and enjoying the mythology around the creation of the rock, when Monica, our young Aussie park ranger, signaled for me to come to an art gallery. She had discovered something—something that blew me away.

The gallery was of indigenous art, and all about one subject. I entered and looked around. It didn't strike me at first until I started reading what the paintings were about, all of them! It was an entire gallery dedicated to the explanation of the creation story of the Aborigine and how they got there. There in front of me was painting after painting showing the story of the Seven Sisters, and the part they played in the creation of their humanism. The paintings spoke of the metaphor of how they had come to be with the tribe, then how they had been chased, and had "fled into the skies." I almost fell over. I sat for a moment and took it all in, for every painting in the gallery showed the seven mythological symbols and the same creation story that Makua had related to me in the North Pacific.

I spoke to the shop clerk, a young tribal member, who told me the whole tribal story as he knew it. Indeed, it was the Pleiadians. He also told me that he had read a book that reported this same story was present for the Greeks, the Africans, and even some in northern Europe. With the specter of 2012 looming large, we now even find out that the Maya had the same story! The seven main temples of Tikal represent the Seven Sisters energy. Here I was in the middle of Australia, in a cultural center honoring the tribes of Aborigines who had been there for tens of thousands of years (their records), telling me about Pleiadian creation for all of humanity.

There are some names of the original elders of this tribe that are so sacred that they cannot be mentioned out loud, and are not even known to the tribal members. Could these be the original Pleiadian teachers? This is my conjecture, and perhaps even me romanticizing what happened at Uluru. But there is acknowledgment within many Aborigines that Uluru is the beginning of their continental Aboriginal civilization, before they moved out and explored the rest of the continent.

What are the odds of this, that so many indigenous from all over the earth would have the same story? They had never met, had no way to communicate, yet they had similar mythology around the seven sisters star system and how creation had begun. It had all begun with sentient Human Beings, who were fully developed and conscious, not developing cave men or savages. All creation stories are that way, even the one in the Garden of Eden. It begins with Humans who are not aware and who then become aware.

A few days later, Kryon gave a very profound channelling for the New Zealand crowd in Auckland. He spoke of the seven sisters of creation and how Lemuria was created in Hawaii to become the oldest, most long-lived, and most harmonious single civilization on the planet. He spoke of outrageous timelines that have no proof in science, but that would eventually be uncovered. He told us that when the waters started to rise (when the ice started melting) that many of the Lemurians of Hawaii became seafaring and followed the currents directly to (you guessed it) New Zealand. He pointed out that although Australia and the Aborigines were far closer, the currents didn't support an easy access to New Zealand, so the Hawaiian Lemurians were the ones who became the indigenous of that South Pacific nation, and became the Maori of New Zealand.

He went on to describe the fact that the indigenous of New Zealand were Polynesian, not Aboriginal, and that although Australia had more than 700 tribal languages developed over time, those Maori tribes in New Zealand did not disseminate nearly to that degree, and to this day this nation is known for the wisdom and harmonious nature of their indigenous people. We also know that their language is very close to the Hawaiian language!

It starts to make sense to those who are esoteric. Still, to those academics who see this, it remains coincidental, and part of "what ancient civilizations tended to do with the stars and constellations over time." I fully understand the logic in this and celebrate the scientists who want to remain detached from such a wild premise. Still, to me there is something here even for science to discover in the future. Most scientists will tell you that too many coincidences eventually beg looking into.

To put an exclamation point on all this, I had a seminar attendee come up to me at the end of the channelling. She seemed to be part Maori, but it was hard to know. There are many families who have integrated with the Maori lineage, and it's very common in New Zealand. She took me aside and gave me some information that showed that, indeed, the Maori had a very similar creation story, although veiled in the names of a joined parent, mother/father sky, and their six children. Their name for the Pleiadians is very similar to the Hawaiian name *Makali'i*. Theirs is *Matariki*, and they have created a New Year harvest festival around them that exists to this day. Here we see very similar scenarios as the Hawaiians and the Aborigines and the Maya—a Seven Sisters creation story. I wasn't surprised, but I got the chills of validation as though Kryon was telling me, "Here it is again for you to hear personally, right where it happened."

So within the teaching of the various layers (energies) of DNA, you will see the "Lemurian" layers. Then you will know the background as to why they are there, for these will be the ones that were placed there, especially for our spiritual lineage, by those brothers and sisters from another star system, viewable to us in both hemispheres of our sky.

It was done on purpose, appropriately, and with full approval by all. We are the ones who planned this earth from the beginning, and who may have actually been part of another "earth" like ours a long time ago in a star system called The Seven Sisters.

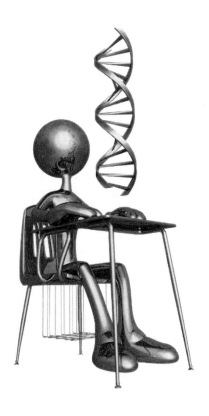

Chapter Five

The Teaching Begins

Kryon

Book 12

Lee Carroll

Chapter Five
The Teaching Begins
by Lee Carroll

LAYERS: From the beginning of Kryon's DNA teaching, he warned and instructed and almost pleaded that we not linearize DNA. We speak of the "layers," yet the very word is a linear word. It conjures up a visualization of a layer cake, one layer on top of the other.

For those who are really not familiar with dimensionality or some of these words, let me give you just a very brief explanation. Something linear is like a step ladder. It has one thing after another. Time is linear, since it is a step ladder of seconds leading to minutes, etc. Time seems to go in a straight line. Counting is also linear, for one number is higher or lower than the next. Math is linear, since it draws logic formulas from a linear system. So "being linear" is actually our very normal state in 3D, and what we are used to.

When we get to nonlinear things, we don't do well, for they seem to have no logic. Can you think of a world without numbers that are higher or lower? How about an existence without the clock? In a nonlinear quantum world, time as we know it does not exist. Instead, there is only one "now." If you absolutely must linearize all of this for your sanity, then think of time in a circle and not as a straight line. Think of numbers all in a jar scrambled up, instead of on a page in a sequence. Think of getting in an elevator and instead of it going up and down, it just "goes" to where something is and lets you out—and it may not move to do it!

If time is in a circle, then it had no beginning or end. Isn't it interesting that we have no trouble with an "endless" existence, going to "the great beyond" as a soul, and being there forever, but

Chapter Five

if someone suggests that there was no beginning to all of this, we just can't grasp that? No beginning? Everything has a beginning! What do you mean?

So the linear mind is biased firmly in 3D. It has no difficulty with things that go "from here to eternity", but can't except anything that might have existed forever in the other direction (the past). Kryon has told us that we are eternal in both directions on the linear time road—always have been and always will be.

According to the statistics, more than 85 percent of humanity has some kind of belief system, and almost all of them believe in the afterlife. Only a small percentage believe in the forelife! Do you see the bias here? Somehow when you are born, your soul also appears as though it didn't exist before, came from nowhere, and never had a prior existence. This just yells "spiritual illogic to me", yet it's the bias that is prevalent in most major spiritual systems on the planet.

Could it be that the way God works is outside of the purview of what we understand, expect, or think is reality? Isn't it interesting that the word *afterlife* is in the dictionary and *forelife* is not? It simply isn't part of our thinking, since we are entrenched in 3D. We feel that a multidimensional state is for theorists, not for us. From what I know and have studied, I think we are going to get a radical awakening about what is "real and not real" regarding the whole quantum issue, and it may even be discovered in my lifetime. Astronomers are buzzing with some of the latest things they are seeing that are only explainable through a multidimensional reality. I believe we are very much part of the quantum reality, but kind of in our own 3D goldfish bowl within it. Thus, we are so smug that we seem to know how our bowl works, and tell everyone that the entire Universe is also a bowl and it works like ours. We then have "bowl science" and a "bowl God".

DNA is interdimensional, or as we might instead say, *multidimensional*. This means we have to look at it and study it all together, and not separate it. Yet we ARE going to separate it, and yes, we are going to call it LAYERS. There are 12 LAYERS. *"Why do we do this,"* you might ask, *"when Kryon pleads with us not to?"* The answer is this: Kryon isn't asking us not to call them layers. He is instead asking us to *understand* that they are not layers, even as we speak the name *layer*—for you and I must exist in our 3D state, and we have to study DNA in this state. So all through this book, think "energies" instead of "layers" when it comes to thinking about this study. That will keep you from asking the obvious linear questions like, "What layer is next to this layer?", or, "If they are stacked differently, what will happen?" Those questions can't be asked, since DNA is not linear.

When I was a young man in school, a buddy of mine was very interested in cars. He took auto shop and he loved it. One day, he told me about the complexity of the automatic transmission. He indicated that when they went to class, the instructor had an automatic transmission still in its case laying on the floor. Then he began to disassemble it. As he went, he admonished the students to understand the "attitude" of the machine. He took one part out and told the students that now the transmission would not work. It needed every part. Not only that, but what the parts did depended on what the other parts around it did. And not only that, but you really couldn't see any of them work when they were apart. He went on, taking parts out and laying them on the floor. He told the students that one part related to the next part only if… and he gave speeds and ratios of gears, fluid mechanics, etc. Each part was "relative" to a situation, unlike a clock where each part always did the same thing in a linear fashion.

When the instructor was finished, there seemed to be hundreds of parts on the floor. This was obviously a non-working transmission that was only there for teaching purposes, for the fluids needed for it to work were not present—another complexity that made the whole thing work or not work. Then the instructor gave them the teaching that he really wanted them to hear. *"You have to study the parts outside of the system to know what they are intended to do inside the system. But none of them will actually 'do' anything without all the others around them. They are dependent on the whole system being assembled, balanced, and aligned."*

We are going to take DNA apart spiritually. We are going to pretend the layers or energies are linear for the purpose of the teaching, even though they are not. We are going to pretend they "do something" as they sit there being examined one by one, but they don't. There will be a tremendous tendency for all of us to see them each as a "whole" layer unit, but they are not.

You will have your favorite layer, but it won't matter. You can't "activate" a piece of DNA (even though Kryon speaks of this in one of his channellings). Instead, you are "addressing an energy group"— one without containment or cellular structure, a concept. You can't even spiritually ask for one layer to be energized over another, for a Human doesn't know the quantum rules of activation, only the parts on the floor. DNA is intelligent, as you will see, and "knows" what you need.

Think of it this way: You are presented with tomato soup. The soup tastes good and it nourishes your body. You love it and ask the cook, *"Please show me the flavor in the soup."* He can't! It's integrated within the soup, not some piece that can be plucked out. You can't pluck out the salt, or even the tomato! It works as a soup and it's now a "system".

DNA, however, is an intelligent soup. Therefore, we must learn what is there, but not be tempted to direct Spirit to do this or that. If we do, nothing happens. Instead, we use our intent to raise our vibrations or become more multidimensional. The DNA reacts by activating the various energies within it which it *knows* we need. This is truly a multidimensional proposal, and not the least bit linear. It affects how we approach it, yet many will be telling you they are activating this or that layer. It's fine. Just go with it.

The layers are numbered, and that's linear, too. Everyone reading right now wants to know the name of each layer and what it does. This is very linear of all of us, and Kryon has done this for us. However, soon you begin to see that it isn't quite that straightforward. Each layer has a sequential number, one through 12. The number of what you think of as a "layer" seems to sit there in a stack with other layers, yet I've already told you that it isn't in a stack at all. Therefore, the numbers become energies with numeric labels and begin to tell a bigger story about the layer/energy you are studying and its relationship to the others.

So again, as we proceed, think of this entire explanation as *DNA esoterically exposed* and laid out on a metaphysical table as a bunch of pieces and parts. See it as a complex engine in which you are studying each section, each piece and each connecting wire. When it's on the table in front of you, it has labels, numbers and instructions for use. But when it's being used, it's put back together in a system where all the parts are melded—no layers, no parts, no labels.

It will again be a propensity for a linear Human to think of Layer One as being first and layer 12 as being last. They are not. They are in a circle on the table—just more metaphors to try and force the linear mind to think in a multidimensional way. However, if you really did put those numbers on a table in a circle, do you realize that it would be difficult to see where the starting point is? Take a look at the illustration at the top of the next page. Immediately go to the starting point. It's not that obvious is it?

Chapter Five

The Great DNA Activation Issue

How many processes, waters, mantras, music, illustrations and seminars have you heard about to "activate parts of your DNA"? By this time, probably hundreds! My spiritual partner, Jan Tober, has a profound, customized process that activates DNA. Using the energy of your name and other information, she produces a customized audio CD by singing and playing beautiful crystal bowls. It's an awesome thing to hear, especially when it's your energy she is producing in this way. (see the color illustrations in the back of the book for more information on this process) So, what is really being activated? The answer is whatever your DNA wants to activate. Does it work? Ask those who've had it done. Yes! Something *is* happening. So what does it all mean?

The real truth is that this term *activation* can mean so many things to so many people. First, DNA is multidimensional, so the layers are not really available as individual objects, like chemistry would be. Yet even Kryon speaks of *activating our DNA* on many levels so that it works better with the body, and in the first of the channellings in this book, Kryon gives a discussion. (Page 222).

So the issue is far more complex that thinking, *"I'm going to go inside and tweak 100 trillion Layer Twos!"* It's kind of like saying, *"I'm going to reach in and grab all the purple in that rainbow."* (This is also from that first channelling.) It just doesn't happen that way. The complexity here is really about us realizing that it doesn't much matter, since DNA is going to do what it does no matter what we call it. So in that Kryon channelling, I mentioned about a group of layers that is being activated together in this new energy; think quantum. Think "energy." After you read the meanings of DNA later in this book, you may understand that it's not about physical layers at all. Rather, it's about concepts and how we are using our intent to bring the concepts to a higher level, thereby letting the DNA select the groupings and energies to work with.

The fact is that there really are good processes and ideas coming forward that are claiming they are "activating" DNA in various ways. The results are really viable, so indeed something beautiful is happening in many of these cases. But Kryon has told me (and will speak of it again later in this book), that what the healer really does is *"increase the efficiency at which DNA communicates to the body."* So the very best a healer can do is to work with that frequency, or vibration, or whatever it actually represents in a multidimensional state.

Think of it as a circle of chicken and egg. The DNA is in the cells, and the cells are "listening" for instructions. The DNA is "listening" to Human consciousness to help tell the cells what to do. So the DNA is constantly "talking" to the cells with instructions for the whole 3D body. So what comes first, the "listening" or the "talking"? For logic might say that if there is nobody listening, DNA could talk all day! The answer is that the consciousness of the Human is the key, and it kick-starts all the DNA layers to do their job better. The talking and listening is, therefore, mixed together in the "soup". However, do you understand that without us "talking" to our cells, that DNA is just resting, doing the automatic things within our body? DNA is always listening, according to Kryon. Therefore, it's our job to inform our own DNA what we need. Again, the key to change in the Human body is *information*, not chemistry.

So what layers are working better? The answer is the ones that are supposed to. You might say, *"Yeah, but I really want to work on Layer Two!"* Actually, we can give our intent for layer two, but the DNA decides what layers are activated, not you or the healer. This is all kind of silly in some ways, for it really doesn't matter what you call it. When you begin to give intent to do anything at all with your DNA, it formats and configures everything for you!

I truly do not wish to offend any healer who is activating DNA, for this work is a wonderful healing modality of high consciousness in a new energy. I just wanted to insert the small voice that says, *"Don't try to count it, name it, or label it,"* for DNA is sacred and you can't really change the multidimensional parts. Instead, the healer helps the Human to activate his own consciousness to allow the DNA to activate itself in the ways it knows are needed. So which layers and energies are activated are the business of DNA, not us.

The very variety of attributes of what is being offered should hint at the complexity, but Kryon says that all of it works! This is because the body and the DNA are fine tuned to "know" what you are doing and thinking. It has been sitting there for eons, waiting for us to work with it and give it information. Now we are, and all of it is food for the new paradigm of DNA processing. We like many, many kinds of food, and we don't really care how the nutrients are allocated once it gets inside the stomach. It all lands in the stomach eventually, and the stomach decides what to do with it, not us. DNA is similar in that it kind of activates itself with the energies that need it, when we start the process of allowing it to do that!

The first Kryon channelling of 2010, given in Colorado, really is about this subject, and it's very informative. I have presented it later in this book. I think it gives a far better summary about activation than I can here. Enjoy it! (page 222)

Numerology

Kryon uses numerology within the revelations of DNA and it's a tough concept to explain. It's tough because you are either one who understands that numbers have energy, or you are not. It's perhaps one of the most esoteric principles in metaphysics, and it's one of the oldest. But it also has a swarm of misinformation around it, coming from a plethora of individuals who have assigned new meanings, have their own systems, and use it for purposes far and above what it historically stands for. Did you know that is it one of the oldest "sciences" in history?

I interrupt this discussion for a discussion. Okay, I do that a lot. When I tell some people about numerology being one of the oldest sciences, they say, *"But Lee, you have to remember that this was back in history with paganism, idol worship, bloodletting, and occult practices."* They object to it even being called "science." Their argument goes like this: *"We are modern spiritual people with more information about how God works than ever before. We now have been given mono-theism, unity thinking, and a modern peek into what God wants from us (usually something to do with a prophet who you need to worship). To look backwards is not correct, for they were savages back then."*

This argument really sounds logical and actually makes sense. But when you start asking, *"What did the ancients actually know?"* you begin to get an uncomfortable feeling that we may have absolutely lost everything that was ever known or given to us and are starting over. Our "modern religion" may be the elementary movements of a baby trying to learn to walk with God, where the ancients knew far more, perhaps even a multidimensional understanding of spiritual energy.

The study of the 2012 Grand Alignment is such a wonderful astronomy lesson! It speaks of a 26,000-year cycle of the earth within the Milky Way (our galaxy) that creates a situation where the wobble of Earth again creates a perception that aligns our sun with the "dark rift," or the galactic equator, as it looks toward the center of the galaxy, in or about 2012. This view, or alignment, with the equator of the galaxy will last for 36 years and then the cycle starts all over again. It's basic astronomy. The thing that is startling is that (gasp) almost all the ancients knew about it, the timing of it, and the wobble of the earth! The Mayans (of course), the Aztecs, Toltecs, Chinese, Egyptians, Druids, Hawaiians, Aborigines, Maori, our own indigenous Indians and on and on—they all knew! We know they knew, since many of their ancient writings speak of it. Now, this means they had to know about Earth's movement within the galaxy. They at least had to know that the earth was round, that it went around the sun, and that the sun was part of a larger group of stars (our Milky Way). This information is absolutely needed for them to have calculated an event that describes our 26,000-year cycle within the stars.

Our history books talk about "modern men" discovering the new science in the 1400s and 1500s. That must have been magical to be part of such a revolution in thinking. Let's see, the earth was flat and Galileo was under house arrest for writing that the earth went around the sun (the Pope didn't like that). The men in ships had to be so brave to venture out knowing they would fall off the edge of the earth! Say what?

That's it, my friends. Think about it. The ancients of more than 4,000 years ago knew about the movement of the galaxy, and thousands of years later, we were pondering if the earth is round? Does it strike you that something may have been lost along the way? It was! And if history is trying to tell us something in this, it might also occur to you that there may have been other things lost, too! What if spirituality, the seed core of our Lemurian quantumness, and even our intuitive relationship to the creator, had been completely lost? The more we study the ancients, the more we realize that they knew far more than we do! No wonder every indigenous culture honors the ancestors so much! They know that much was lost.

Therefore, there is the possibility that the systems of the ancients used advanced ideas, and numerology was one of them. It truly is a multidimensional system, since it deals with concepts that are outside of 3D and sometimes eye-rolling to "modern thinkers." I still contend that we are in the "era of spiritual ignorance, of lost truth " and that someday all this will be called the "spiritual dark ages." After all, these days the truth of God is obscured in Human bias, where wars are fought about who loves God the most, who has the *correct* system, and who has the *correct* prophet—all in a singular bias as though God was somehow in some Human box, captive to lower Human 3D ideas, and only available to one group of Humans—the ones who owned that particular box! *"My God is the one true one, and loves me the best, and because you don't believe me, I'm going to have to kill you, because God just doesn't want you around thinking like you do."*

This entire scenario is grossly ignorant of the idea of a huge, loving God who created each of us in the image of divinity, and who doesn't care how we contact the "prophet inside," or how we find the Higher-Self. For each Human has his own path, his own way of worship, and his own idea of how to love that part of his reality that is God. And when it's all over, and in death we present ourselves back to the place where we came from, that loving creator won't object to the color of the car we arrive in (a metaphor that Kryon has given over and over).

To think otherwise places God squarely within the purview of 3D Human ideas and thought—one of reward and punishment, revenge and anger, angelic wars, just like the Greeks did with their mythological men-gods of old. If I was born in His image, then I'm magnificent, not dirty, and not limited to climbing one "correct ladder" to accept some man-made doctrine that will cloak me in appropriateness to be accepted into the "club" when I die. Instead, I will simply return home, having the same party all the others will have, received with honor and glory and thanks for holding the light on the planet.

Numerology is a multidimensional system of old, and a conceptual system of energy developed by a far more advanced group of spiritual thinkers than we currently are experiencing. Within the pages of this book, the system I'm going to speak of is simple and right out of ancient Tibet.

The Energy Around Numbers

First, a basic discussion of the energy of numbers: Why would something as simple as a number have energy? For many situations, many reasons—for instance, if you have a number *nine* just sitting at the bottom of a page, it's very different than the word *of* sitting alone, for instance. The *nine* has a message that the word *of* does not. The word *of* tells you nothing by itself, but the *nine* at the bottom of a page tells you that there are *eight other pages* potentially created before it. It, therefore, has a logical message for you, and it's just a number.

What just happened is that if you look at the number *nine* on the bottom of a page, it has a message. It means it's the *ninth page*. It points to a situation that tells you that somewhere there are eight others. So, did it have energy? Yes! So what about a number alone, not at the bottom of a page? It's just ink on a page, isn't it? But ink on a page is also scripture and love letters and many other things that change us. But a number? Consider this: What if that number is actually a name of someone you know, or perhaps an angelic someone? Would that change the situation? Well, it's the best I can do to give you the reasons why numbers have energy and meaning. They do have "angelic names" to me, and they often give entirely new interpretations to situations we think we know all about. Think of them as abbreviated energetic messages contained simply as numbers.

Numbers are part of an ancient system of meanings, and they fit together in a beautiful way to give alternate perspectives. In ancient times, the ones who "threw the bones" or "read the tea leaves" were doing the same thing, often counting the points of the bones that faced a certain way or counting the leaves, etc. It probably all centered around some kind of numerological patterns and was generated by what appeared to be a random event (i.e., the toss).

Numerologists are often employed to interpret the numbers around events or names, as a way to "see" more information around a situation. Sometimes they are commissioned to change the "energy of a name." Let me tell you how that might work and why.

However, I'm going to stop the discussion yet again, for many people ask me about the word "energy." What does it mean? For those in metaphysics, it is used a lot. There is energy around situations, people, systems, etc. What is it? Here is the definition from Charles Fillmore's *Metaphysical Dictionary: Energy: Internal or inherent power, as of the mind; capacity of acting, or producing an effect.* By the way, who was Charles Fillmore? He was the founder of Unity church and also (ready?) a channeller!

Energy, as we use the term, is something we can sense or feel around anything, anything. It does not have to be physical, either, as in the energy around a situation. It might be drama, love, the "elephant in the room," sorrow or grief, or exultation. It might be irony, frustration, or celebration. There is energy there! Don't believe it? Then the next time you laugh or cry in a movie, just deny that there was any energy there that made you do that. It's just a moving image with sound, after all. But the situational setup is almost palatable, bringing to life the incredible energy around Human situations and compassion. We are "tuned in" to it and react.

Ever sit down in a movie theater next to a stranger, then you move? What just happened? Call it intuition if you wish, but that person may be mad, sick, upset, or perhaps "the energy just doesn't feel right" (smile). Most of us feel energy all the time, but we don't label it. Most people who are not in metaphysics never even use the word. Instead, *it just didn't feel right*, or *gut instinct* said this or that. What they are doing is giving you their word for *energy*.

Ok, energy around emotion is obvious, but how about "energy around things." This is more of a stretch to those who have picked up this book by accident in a metaphysical book store. Can a thing have energy? How about a crystal? Of course. Many can actually feel this with their hands. How about something you love, such as a photo of your mom or dad or your child? Yes! So what is happening is that there is a combination of things taking place here that create energy. Sometimes, like in the crystal, it's almost totally contained by the object. Other times, like the photo, it's about your memory and the situation around the photo. But in both cases, there is energy around a "thing" created by a vast diversity of attributes.

This is the study of energy around numbers, and it's just about the most complex thing in metaphysics—even more diverse than astrology and the many kinds of interpretations around the movement of the planets.

The Forms of Numerology

Here is where it gets dicey, for those reading this book who have been using numerology as their livelihood for years will probably be offended by the simplicity of these explanations. So I'm going to give a disclaimer! It's really only so the pros will understand what's really going on in this discussion.

There are just dozens and dozens of kinds of numerology. Those who really don't know about it might say, "*Well, which one is correct?*" The answer is, all of them! How can that be? Because the variety of the systems represents what they are trying to achieve. If you just came from another planet and saw that we have thousands of kinds of food, would you ask which kind was correct? The answer to that question would be, "the kind of food the Human body needs." The same is true in numerology, for different systems actually produce better kinds of answers depending on what the numerologist is trying to "see" in the system for you.

Let me give you a glimpse of the complexity. (Warning, this is not easy to understand.) Numbers have energy, but so do their relationship with other numbers. So if you are really going to do this right, you should study what the relationships between numbers are all about—not just the math, but the energy relationships as well, like the circle of nines, the Golden Ratio, etc. There is super-simple, linear-based numerology, which is a nice beginning way to study it, all the way up to multidimensional numerology that uses "influential numbers."

One of the recent channellings of Kryon (not in this book) highlighted this quantumness. He spoke about a number just "sitting there" minding its own business, but it "missed" the ones that used to be next to it, since they affected one another and changed the energy around each. This is a metaphor. Don't send me letters about how silly this is and that numbers are not like pets and don't miss their friends. It's just an analogy.

Even in a linear system, if it was a *two*, for instance, you might say it has different energy than just a *two* because it also has the influential energy of the *one* on one side and the *three* on the other (in linear counting). Now, if you know the energy of those other numbers, you can see that they might influence the *two*, depending on the circumstances surrounding the reading and the common mathematical relationships between numbers in general. So, a *two* is modified by circumstances around it. Wild, huh?

Here is where it gets really complex. If the *two* was in a formula, or part of the result of a calculation, or a reverse number system, or a series of numbers that represented another system, then the numbers sitting next to the *two* would be different than in a linear counting system. What if it were in a table with numbers all around it? Say what? I'll show you when we do the simple numerology around my name, coming up.

In a moment we're going to look at super-simple linear numerology and find out what that system does with a common name. But first we will assign very straightforward meanings to the numbers one through nine. Let's call these meanings "the number meanings for this book" so I don't run into trouble with those who have older or newer systems. This means that these attributes will fit into the numbers that Kryon gives for the layers and also how they interact. It doesn't mean that I'm giving you some empirical definitions that you must see as "the only ones." They are just the ones for this book.

Again, I have had help from Dr. Barbra Dillenger here, for she has been dealing with numerological attributes in her work for years, and has studied the basic Tibet systems to arrive at very distilled and accurate energy definitions, even for a simple linear system. So these really are ones that have been carefully researched by her, and they are ancient indeed. These are also the ones that Kryon uses for this study, and I don't think it's a coincidence. I believe that there is synchronicity here, and it helps me to describe to you better what the energies are around the layers of DNA. So to be clear, these are beginning simple energy assignments to assist those unfamiliar with numerology.

I want to let the reader know that Dr. Dillenger uses a far more complex relational flow system in her professional work. It's quantum-based numerology from her Tibet studies, and is not the simple linear system we are using in this book. But she has helped with the basics for us here, and with this very explanation, so that we can have something that is at least exposed correctly and with integrity, to those who have no idea about the subject.

Here are the simple meanings of single-digit linear numerology for this book. If you truly understand what I'm teaching here, you will understand that there are no good or bad numbers! They relate only to your situation and your own path. Each one can be wonderful or challenging, depending on the situation it is presented within. This in itself should tell you just how complex and multidimensional this all is. Kryon uses them in a quantum way, which simply means that they are all sensitive to their potentials. But the basic meanings are the ones that follow:

One (1): New Beginnings
A number about "self." Some also see it as a unity number.

Two (2): This is The Duality Number
Human vs. Divine—Polarity.

Three (3): This is a Catalyst Number
Creative—Also joy and inner child energy—powerful!

Four (4): Mother Earth Energy
The physical world—Understanding or being based in structure.

Five (5): Change
A number ruled by Mars in Astrology—a very fast number.

Six (6): Sacred
Communications—Harmony—Balance—Love.

Seven (7): Divinity
Wholeness—Perfection—Learning Life.

Eight (8): This is The Practical Number
Structure—Practical—Manifestation.

Nine (9): This is the Completion Number
Completion—Sensitive and Psychic—Humanitarianism.

MASTER NUMBERS
Eleven (11): Illumination
Twenty Two (22): Master Builder—cosmic law
Thirty Three (33): Christ Energy (Christ as a title, not a name)
(44) through (99): Has not been given, sine we don't have the understanding yet of these more quantum master numbers.

As you look at these, you wonder how something so simple can be so complex that a science could be created around it. Believe me, it's not simple! Again I mention that if you want an eye-opener, consult a world-class numerologist and ask for a reading. You will go out amazed and appreciative of the knowledge that it takes to be a numerologist.

In 3D, the numbers and meanings are kind of fun to examine. They may remind you of people or things or situations. The *one* is interesting. How can *new beginnings* also be about *self*? I think it's a qualification that the new beginnings are always about self… you making decisions, not a situation. It personalizes it.

Two is a number that I have always looked at as difficult, for duality is difficult! But what if it were about mastery of it? It goes both ways.

The *three* is the most exciting number for me. A catalyst is an energy that creates or speed up a reaction when two unrelated things come together, without necessarily changing itself. In situations, it is an energy that precipitates an event. It also seems to involve joy and the inner child. Perhaps Kryon will elaborate on that.

The *four* is interesting. Most tree huggers I know are fours! Honestly, this stuff really starts to make sense when you continue to see it in the real world. I laugh, as the 4-H club at school was about farming and nature things. It's a big organization still around today. The four is about mother earth, and the "H"? It's a double 4 (in our linear alphabetic system). Farm out!

The *five* is well known: Change. Sometimes it's good news; other times it's difficult. If it's getting out of a job you hate, it's wonderful! If it's about getting you out of a job you love, it can be a challenge for awhile. Again, the numbers only reflect the energy of the situation. Many times, "difficult change" turns into wonderful manifestation. It's just that it's hard to see what's coming when God doesn't give you a map. Numerological energy is not static. The interpretation of the number meanings for you always changes, as you do.

Six was not something I really knew about or understood. Kryon has made it a cornerstone, and even has given some profound information about the way it has been misused and obviously manipulated. But it's one of the best numbers around (my bias), and right in the middle of the "hub" of the 12. It represents sacredness, harmony and balance.

Seven has always been thought of as divine, and our own cultural Bible tells us that as well, giving many of the lists of things in a seven arrangement. It is also used this way in many of the other scriptures, such as the Qur'an. In fact, the number *seven* is very significant to many Muslims.

Responsibility seems to be a word you would not want coming up in a reading. But the number *eight* isn't your mother telling you about cleaning up your room and being responsible for your mess. It's about practical energy and structure, and (ready?) it's the manifestation number!

Finally, the *nine* is completion. That's finishing or shutting down, or completing. It could be a phase in your life, a situation, the energy around anything that you are finishing.

Many people even do the numerology of their address in order to know more about the energy of "them in that house." A *nine* number may mean that it's the last house, or the one where they will finish their life's lesson, etc. This is where a professional reader is needed.

The *master numbers* speak for themselves. The only odd thing is that there are obviously nine master numbers, but only three are clearly defined in ancient history, or traditionally used. The reason? We don't have understanding of what they would mean. Wow! If I'm reading that correctly, we are only a third of the way to understanding the energies around us, and it's only *three* are understandable to us, it kind of like a "wink" that we are still in 3D, I think.

Let's do an example...

We will do super simplistic linear numerology on two names. My name is Lee. (It's actually LeRoy, but don't tell anyone, for I want that to remain a secret only to the 100,000 or so who will read this book.) The other name will be someone I don't know: Mel.

Numerology, like so many of the esoteric sciences, takes synchronicity as part of the cycle of who you are, and what is happening to you. Do you understand that? However, most systems are absolute. That's why you call them systems. You have to do this and that, and don't deviate or you will mess it up! It's like what we expect from a recipe. Don't follow the system (the food choices and mixtures) and you'll get something that may make your relatives sick at Christmas.

(Some people may want that recipe!) Or in metaphysics, *"Oh my God, I faced north instead of south when I meditated! Now God won't listen to me for a month."* You think I'm kidding, don't you? I've seen that exact thing in metaphysical circles.

Good linear systems, and there are many, should be dealt with that way. They are 3D. Sometimes a good metaphysical system is created in a very linear way, to enable someone to see a bigger picture outside of 3D, or to accomplish self-awareness and growth. This is true for some of my colleagues in New Age work. I have friends in Europe, for instance, who realize that some things just have to be taught in steps, since that is what makes more sense to a European culture. In the states, however, it's not a step process, as that just doesn't work for most Americans. So who is right in the "steps argument"? The answer is that they both are. So the teaching suits the cultural situation, but creates a win-win for both. Again, we climb the same tree in many ways, all to arrive at the top, knowing the same thing.

Numerology is an ancient esoteric system where there is credibility to something nobody wants to talk about—an attribute where the system "knows" what's going on. Therefore, what seems like chance to you is really only an attribute of quantumness, where linear logic suffers but *potentials* are king.

"There are no accidents," we are told, only potential elements of what you have created through your energy. *"Yeah, but the other guy did it to me! I didn't create it? I'm a victim of it."* I hear that all the time. You should know that the *whole* is seen by Spirit, not just your "accident." You put yourself in the path to be "done unto." You attracted the energy of it and walked right in. It's a *partnership of energy,* and almost everything that happens to us is this way—a circle or confluence of potential energies all mixed together that align to create our reality.

Chapter Five

So what I'm saying is that what appears as randomness has purpose and meaning, for it was YOU and your energy that "picked the card" or spun the wheel of fortune. This is so multidimensional! Yet many see it as ultimate foolishness. In numerology, the way your name is spelled has the energy of your birth around it, and many other things as well. YOU didn't name yourself. Someone else did. Therefore, it has their energy also. In our culture, the way to calculate the "number" associated with your name is also questionable to many, for it's often based on the English alphabet.

"*Wait a minute!*" some say. "*That's crazy. The English alphabet was not around when the sophistication of numerology was created. So how can you use it to interface with this so-called ancient system?*" This is where we lose many people who just can't think beyond the 3D they were born in. The system *knows*. Remember? It's also a current measurement, even though the process is ancient. Why is this difficult to understand? Should we stop using prayer since it was conceived long before the kinds of situations we find ourselves in? Can we pray in English? After all, that language wasn't around when prayer was initiated. See what I mean? These are foolish questions about a *knowing* God, and a *knowing* multidimensional system of reality.

The real *system* is the energy of YOU, of your culture, of your alphabet, of the choices made by those within your culture. So numerology becomes compliant to meet the situation at hand, no matter who you are or where you are. So perhaps you're an American (let's say), born in a place that uses English. This becomes the core of the seeming randomness that attaches itself to you, and numerology works with it.

Think for a moment: Do you think God is somewhere in a spiritual closet when you pray? Do you think for a moment that God isn't aware of who you are and where you are? My belief has the creator with me all the time! This means I don't have to explain who I am, or what a Human is, or anything like that. I'm being facetious here, but some actually run their lives this way. Sunday is the only day God is around, and the rest of the time these people think they are "invisible" to the creator. The reason I'm speaking of this should be obvious. Spirit (God—creative force—innate) is with us all the time. We are a big part of the spiritual system at hand, and we are loved and recognized as part of creation.

My God knows who I am, where I am, my name, my energy, and what I'm going to say as I say it. My God knows my culture, my issues, my Akashic lessons, and what my propensities are in my "randomness." So what I'm saying is that numerology is all part of this "God system." It makes numerology more than a system. It's part of the quantumness of Spirit, and a core element of how much we are understood and loved by the very energy of creation.

In this very simple linear system, we are going to assign a single digit numerical value to the name "Lee." We use the English alphabet with 26 letters, A through Z. A equals 1 and Z equals 26. All the other letters have their values in between. When we spell a name, we will then add the number equivalents up and reduce the number (if needed) until we get a single number that we need to work with. The basic numerology we are going to use only works with *one through nine* numbers, and a few *master numbers*. So it's fairly easy to look at the core meanings. The idea is to get a single digit, then see the energy around it based on the chart I gave you. It will tell us something, even in this very simple linear example.

Before we go further, we halt again for those to be heard who are having trouble with this concept. *"What if I do your name in Latvian? Latvian has 33 letters, so the numbers of the letters might be different. It would then change the final number. This is foolishness since the 'final number' would always be different for each language."* The

Chapter Five

answer to this honors synchronicity and shows the basis for the way it is misunderstood. You were NOT born in Latvia, and your mom didn't use that alphabet when she named you. Understand? So numerology uses the synchronicity of the reality of your personal circumstances in order to work. It *"knows."*

"Yes, but I am Latvian and moved to the USA when I was four! What about that?" My answer is this: Do you really think that this is a secret to the Universe? In other words, all of that is taken into consideration with a system that "knows." If it's a system that *knows*, then it's not separate from the other systems, either. Astrology is very related to numerology, and there is a marvelous, complex interface between them that Dr. Dillenger understands and uses in her work. I'm not even going to go there. It's elegant and out of the box of where you might take it. It's multidimensional. So you can't "trick" the system by accident or on purpose, or wonder if it's accurate simply because your situation isn't normal. The bottom line is that when you see it work, it's amazing what it *knows*. Do a reading with a world-class numerological pro before you decide this is not working for you. You will walk away changed, and you will wonder how a system of numbers associated with you and your life could "peg" you so completely.

As we do two names, remember that this is the simple linear stuff, just for an example in this book: My name is Lee: L=12 (12th letter of the alphabet), E=5, then another E=5. Now, stop everything. I'm telling you that (for lesson purposes) my real name, the one my mom gave me, was LeRoy. Shouldn't we be using that? The answer again is about synchronicity. What is the name I am called in life? It's Lee. The name LeRoy was only uttered out loud when my grandfather was around, since it was his name! (I was named after him.) Everyone else hated it as a name for a little kid. Therefore, *Lee* became the name I am known by, as well as the name associated to my energy. See how it works? For more than half a century, my body, my energy was called "Lee," not LeRoy. The system *knows* this.

Using that alphabet system I described, my name adds up to 22 (12+5+5). Now, I chose my name for this example since I already know the attributes of it, and also since there is a rule in numerology that when you get a double number, like 11, 22, 33, 44, etc., this attribute is treated as a "master number." These master numbers have energies and meanings of their own, so you don't reduce them to a single digit. So my name is a "22." I'll leave it at that, for you can see what 22 means in the chart on pages 86 and 87.

Stop the discussion again! We now return to the former discussion of how numbers have relationships with other numbers. Let's add a layer of complexity here. Even in this super simple system of assigning numbers to letters in your name, you must ask, *"What is the relationship of the numbers next to each other, for they now relate to letters in an alphabetic system, not numbers in a linear system?"* They relate to A, B, C, etc., not one, two, three. So, for instance, the Lee name has an L, which is a 12, next to an E, which is a 5. Hey, 12, 5 is not a linear progression! So we go back to the numbers "missing" or affecting each other (former discussion). The 12 (which is really a three when you add it up) has been sitting next to a five all of Lee's life. In complex numerology, this is important. Is there an underlying eight somewhere in there? (Smile) It's energy, folks, not necessarily math. So how do you ever figure this out? Leave it to those who have studied it all their lives! Just understand there is a whole lot more to this than you think to these systems.

So, do you now see how even a simple linear system becomes nonlinear as soon as you assign linear numbers to an alphabet? The energy of the numbers next to one another really do have an effect on one another. And in a name, it's not a linear progression. Instead, it's letters in another entire kind of system (alphabet). I'll only mention this next item to make you sigh. In a true multidimensional numerological system, the reader also takes into consideration what is "above" the letter in the alphabet. Don't ask.

Let's do one more name...

Mel is M=13, E=5, L=12, so Mel is 13+5+12. This equals 12. Twelve is two digit number and so we have to add the numbers together again. Aha! *Three* is the number for Mel. As we shall see, *three* is a powerful number. But perhaps Mel would rather be a non-smiling, surly monk! (I met those guys at the Church of the Holy Sepulcher in Israel.) His name is getting in the way of that. It's a bad number for a surly monk, since it's about catalytic energy and requires a social interchange. He might wish to respell it, perhaps adding another letter? Mell. This makes him a 42 (added together which makes a six). Six is wonderful for a monk of any kind!

As funny as this sounds, this name respelling is done all the time. Many do it around their pets as well. Now, it's a lot more complex than this, and I'm really simplifying all of it, but this is why you would want to consult a real numerologist, for they would have the meanings behind the meanings and would know the system it came from and the history of how it is being used. These things are all important to many people. So if you meet someone named Leigh or Salleee or Frannk, you just might know what has happened. That or their parents were visiting an alternate reality when their child was born, such as in *Moon Unit* (daughter of famous musician Frank Zappa).

So why all the numerology talk? Because I'm convinced that Kryon took advantage of our DNA "layer" numbering to give us hints as he presented the layers one through 12. Again, they are not in layers and they are not in a stack. They are energies that are multidimensional, so Kryon has assigned numbers to them, and I'm just presenting them in "our linear" expected order (one through 12).

They are not in any kind of importance order, but rather given in a relationship order to enhance the message of energy of what they really are about. So with each layer Kryon presents here, there will be a very brief numerology given around the number, and what it means in the big picture. All this is to get you prepared for a very special discussion that is very sacred and will be presented in a moment.

SUMMARY

Finally, regarding numerology, I give you this: Back in 1989, Kryon gave us the meanings of two numbers. Little did we know they would actually define our age. He spent time on them—the only two numbers Kryon ever described this way in any book. They were the number **NINE** (Kryon, Book One, pages 24 and 74) and **ELEVEN** (Kryon, Book One, pages 15 and 85). Was this a hidden prophesy? I feel it was just a description of the numbers that would mean so much to us all these years later. When we put them together, we get 9-11. What is the energy? It's the completion number next to a master number of illumination. The very event of 9-11 has redefined the paradigm of our age, increased the polarity, exposed the duality, and brought us into world alignment in a way that is creating a battle between the old and new energy, with Israel right in the middle.

So what is the *message* of those numbers together? Consider this: It's the completion of an old age, old energy, and old paradigms, and the jumping off point to an illuminated earth. This number combo was right there given to us by Kryon in 1989, but only in 2001 did it actually mean something to us. Think of the energy of 9-11… just ink on a page, but containing a vast amount of information to most of us.

Nowhere on the radar screen of Nostradamus was the 9-11 event. One of the most world-changing events that will happen in our age, and he didn't mention it. Kryon didn't either, but he clearly gave us the numbers for it right from the start, within the first year of his communication in 1989. We have shifted the timeline of history, the reality, and our own paths.

Another interesting number sequence is one that is the nickname for the 1987 Harmonic Convergence. It is 11:11. If you do simple numerology on 1987, it becomes a sacred energy (the number seven). And if you know what it means to the New Age, this 1987 year, indeed, was the beginning of the shift we are seeing on this planet.

One of the true visionaries of our time is a woman who goes by the name Solara. She facilitated the 11:11 Planetary Activation in 1992 in which well over 144,000 people participated worldwide. She describes herself as an *"intrepid explorer into the Unknown, going where few have dared go before... into the undiscovered subtle realms of the Invisible."* Whatever this means to you, she is the one who named the Harmonic Convergence, *the 11:11*, then did a planetary event around it. I honor her.

Many see the 11:11 number on their digital clocks—way too often for it to be random. I think it's an "energy wink" from Spirit. If you look at what 11:11 might mean in numerology, it is two master numbers side by side. Therefore, it is "Illumination: illumination." This is, indeed, the energy of this shift we are in, for we call ourselves Lightworkers, in an illuminated countenance, started at the 1987 Harmonic Convergence, nicknamed 11:11.

Perhaps we should now speak of the profundity of the power that's in our DNA, for we will need to hear it a few times before it sinks in. By the way, now do our own exercise. Do the numerology on DNA. Then check what it means (smile). Is something new beginning here?

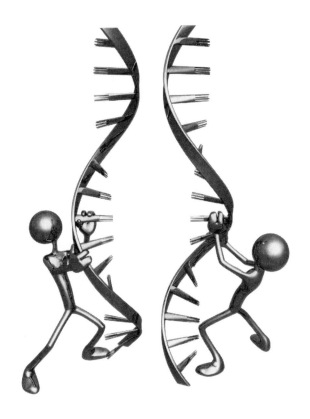

Chapter Six

The Twelve Layers of DNA
Kryon
Book 12

Chapter Six
The Twelve Layers of DNA
by Kryon

It's time to list the layers and their attributes. However, I'm going to do something that hasn't really been done a lot in my work up to now. I'm going to let Kryon take over and do these DNA explanation sections. It will be more powerful that way, and far more complete.

Most of the Kryon books are made up of transcriptions of live channellings from around the world. Very few are written originally on the page. Only Book One was done that way, before I was channelling live. So it's time to return to my roots and let Kryon give you these energies as he originally did with me over the five years that it took to slowly unravel all that is now ready to be given to us.

Lee Carroll

Greetings dear ones, I am Kryon of Magnetic Service.

Blessed is the Human who is interested enough in his path to have his eyes on this page, for this message is one that I saw the potential of many years ago. To me it had already been written, already been published, representing the book you hold in your hand. But my partner was in his "infancy of spiritual growth" and had no idea about what was before him. Now he does.

Let me take you somewhere. Pretend for a moment that the very essence of your being could be viewed in a way that is new

Chapter Six

and marvelous. Let me take you to a place that has the very fabric of the secrets of life itself. This is a place where dimensions cross dimensions. If you had multidimensional sight, you could clearly see it. Like the magnetic lines of flux suddenly being revealed to the Human eye, the colors would dance like the light reflects off of slow stirring waters. There is song there! For dimensional overlaps do two things; they sing and they make light. The photons you can measure in 3D, but the songs are the kind that are sung at the edge of the world where the solar wind meets the magnetic grid, for the same kinds of processes are at work within the DNA.

Let me take you to the interior of DNA itself and reveal to you the esoterics and the love that this process has within it. For more than chemistry, this *DNA event* defines the core of sacred life, the love of God within the Universe, mixed with dimensional confluences and the joy of creation. DNA is the crossroads of God and man, the mixture of quantum and non-quantum, and it vibrates with the essence of the truth of the Universe. If you could sit within the double helix and observe all the vibrations as I can, you would be in awe. For within that 3D structure surges the history of the Universe, of mankind, of the seed race and their love for you, and your relationships to the ages and to the earth. The multidimensional light show is grander than any that any Human can conceive, for this is the kind of light that is not seen, but felt. It sings a sonata of melding energies that soothes the soul, and the strains are similar to those you "hear" when you are on my side of the veil. For DNA contains the creator energy, your energy, the Human transformation energy and that of all your lifetimes.

Your angelic name is sung on top of it all and you will soon behold what is taking place—a celebration of the royalty within, for all the chemistry, both quantum and non-quantum, bow to the whole and line up ready for the crowned one to speak his or her instructions [the Human Being]. It's a place where physics meets spirituality, and a place where complete peace and solace of consciousness are achieved. The bridge to the creator's reality is there, and in each DNA molecule there is a mini-portal that leads to a multidimensional Universe.

This thing you call the double helix is sacred, unique, and is that way only for the Human. For the DNA of other life does not have the creator inside, but rather it is designed to *see* other DNA that has creator attributes. Said simply, it means that anything with DNA on this planet *knows* about the Human creator DNA, and bows appropriately to it. Even a vegetable *knows* who you are and lives to nourish the life around it and you. At the basic DNA level, animals are also aware of why they are here, and how the Human carries the creator energy on earth, and that they are only here to support it and the planet.

There is spectacular beauty in both the view at hand and the purpose of DNA. For although it exists in 3D, it is one of the only structures that is mostly multidimensional, but hiding within a 3D chemical shell. Sit here with me for a moment and think of what it represents. The atomic structure of the Universe also sings with creation, but it represents the building blocks of everything, made up of multidimensional parts that Humans can only see a portion of, since Humans are in their own dimensional reality [3D], for the rest just look like space and emptiness to them. But in the space between the nucleus and the electron haze is the "soup" of creation, the multidimensional glue that sets the rules for the way complete physics works. Also in that glue is a natural nature bias, for it is designed to create life, over and over.

At the atomic level, the smallest of the small, there is a master plan, and it is not random. For it will configure itself over and over to let life start everywhere! This means that life itself is sacred, just

Chapter Six

as you might imagine, and as your intuition tells you. Now you can see that the DNA molecule is the result of this design, and is not random at all. Perhaps it's time to give it the respect that only a quantum thinker can? Can you imagine what it carries with it in order for it to create the mastery that is available to you?

Sit with me. How old do you think the songs are that you are "hearing" within the complex intertwining of multidimensional energies? Think of a time long before the earth was formed, all contained within this small multidimensional double helix. There are more than 100 trillion others exactly like the one you sit within, and they all can speak as one, which gives them better communication skills than any existing process on the planet. DNA memory speaks of watching the earth form and all the life coming and going. It's all there, like a recording device in your technology, only this is multidimensional, so there is no space limitation. The mini-portal within DNA is the key, for it connects the Human Being with all that is and ever was. Therefore, the DNA becomes a portal of connection with Spirit itself, and is never without this connection.

Yet there is no boasting of power or energy or purpose. Here it all sits, vibrating to a master composer, who is the unique Human Being it represents. There is simply no other Human who has the exact DNA. Even identical twins don't have identical DNA. Only four percent is identical, and the rest is unstructured and multidimensional, speaking of the individual lifetimes and experiences each has had in the historic footprint of the earth and the Milky Way.

You might sit and ask, *"Why do I feel so much love in all this, for DNA is a blueprint and an attribute of biological creation? Why is there such an amazing feeling of love? Why does it make me ponder other things? What is here that would create such a grand feeling of belonging?"* The answer might surprise you, for DNA is the mastery substance of creation on the planet. It, indeed, carries the image of the creator, and reflects it in ever way. It is elegant in the way it delivers itself to the Human body, and yet it has a high consciousness that hides, ready to be understood by any Human who wishes to examine what makes consciousness happen at all.

It carries with it all the history of creation, including the galaxy you sit within. It is the glue that is ready to be placed into the crevices of the breaking Human heart. It is the peace that passes all understanding. It is DNA that is responsible for consciousness, not the Human brain or gene set, for the brain is simply an organ in the head with synapses, and the genes are chemical engines. But DNA creates the *informational instruction signals* that keep the body going and the intuition that says, *"Yes, there is a God."* Each strand is identical, each one belonging to one individual soul who is visiting the earth yet again for a seeming moment of time. DNA represents the energy of multidimensional information. It is not the chemistry that creates consciousness, but rather it is information, and always has been.

DNA has various degrees of efficiency, and depending on the vibration of the planet, these efficiencies are produced at birth. At the moment, the Human race is at an efficiency of approximately 30% activation. This is a linear concept, but it is all we can give you that will help you to understand how things are working. The various masters of the ages had 100%, and you can see what they were able to do. First, each one had such a connection to God that they were considered God themselves. Each one had a peace that made Humans want to be around them, fall in love with them, and even worship them in death. This is the mastery that is inside the DNA, and these masters who walked the earth each came to show you what is possible. Each of them carried informational messages of love, unity, and peace. Some of them showed their power over matter, over life, over the elements of the planet. These historical reports are not exaggerations, for this is, indeed, what took place. For when the DNA is working at 100%, you have the empowerment of the creator fully manifested within the Human Being. The efficiency factor is not chemical, but informational. It's about multidimensional energy… not chemistry.

A master can choose the time of his ascension, the time of his birth, and even who he will be when he comes back. A master can pull upon the Akash of his own DNA to bring forward attributes from past lives, both consciousness and biology.

Sit with me and see this, for it's in you and every Human around you. The entire purpose of the shift at hand, and this book in particular, is to enhance your understanding of the power you have within yourself and reveal to you that there is an amazing process outside of your 3D logic that is within you. Can you increase the percentage? Yes. Is it time? Yes.

This is the entire reason that Kryon is here.

About the Book

All along, this publication was seen as the 12th one [the 12th Kryon book], just like the number of energetic DNA layers. The 12, as you have been shown, is really a *number three*. This is the catalyst number, the one that "does something with something else." Don't you find it appropriate that we begin a teaching with that energy in 2010? [Kryon smile] For this information has the ability to resound within the pieces and parts of your quantum state. There is a remembrance for you just laying on the surface, asking you to prod it. Each Human has this, for it is the thing that creates faith—trust in the unseen, a very illogical, 3D attribute. The "threes" are around you here. Find them. For my partner, I list them: This is his 21st year channelling Kryon, the 12th book in 2010. There are *three number threes* there. Add them up for a nine—completion. It is time to reveal the DNA.

"I just know it's true," said the Human when asked, "How do you know God loves you?" There are certain innate energies that are the very substance of what holds you together. Almost the entire Human race searches for the creator as soon as they are old enough to ponder their own existence. Now it's time to peek into why that might be and the profound sacred energies that every one of you has.

After this explanation, we present a channelling given in a place that most of you reading here will never go. It was planned that way, and in that channelling, I told my partner about it—that it would appear here [in this DNA book]. Riga in Latvia is experiencing an awakening as you read these words. There is economic darkness [2010] upon them, yet they are still filled with the hope and joy of a new beginning. It was the perfect energy to give them a profound message about exactly which of the energies of DNA were being activated in humanity at the moment. In another place, it would not have had such meaning, for the Latvians heard it and knew what it meant. I told them then that it "was not a western message." In other words, it was for everyone, but given to them first because of their full understanding of what it means to come from nothing and begin all over again. In that arena there was silence, and a great, deep reaction to what they *knew* to be the truth. I chose Latvia for this in order for those in the western world who will read this book first to understand yet again that this is a humanity message, not one for just one culture.

Your cells are designed to *ring* [vibrate in recognition] when they realize truth is upon them. You can actually feel it and know it is so. It is the truth of information, not chemistry. Some of you have the chills or other physical attributes as the whole of DNA tries to give your brain an indication that there is a "whole" message being received that is not necessarily the one you are hearing. It is presented in "the third language" [another three], which we have identified as catalytic in nature; it changes something. Now you may understand this, due to the simplified explanation from my partner about the meaning of the numbers. You see, there's more than one way to receive information within Human communication. All along, the very cells receive energy in every situation. Often, you feel it as fear for no reason, or joy for no reason or "feeling" to go do something. Think of it as a second brain, perhaps a spiritual one, that does not use linear language at all. It's a quantum effect, so you can't explain it. But it's there and ready to be called on as we move forward here in this book. Some even call it a "gut instinct."

So be still for a moment. The book can wait. You may wish to ask Spirit, your Innate or whatever you think is the essence of creation, *"Dear Spirit, let me 'feel into' this message and have my own creator energy validate this truth."* For when you begin to do that, all that I'm about to present won't be new at all. Instead, it will be something you can say, *"I knew that!"* At that point, dear one, we are all *one* with this information, and you are not a student at all, but an observer who is blessing the information and celebrating what it means for you and those around you. The mastery within you is active, and you are beginning to open up to multidimensional communication. You will also realize that this message is one you have been waiting for and that it could not have been given before now, for humanity simply was not ready. Now it is.

The Organization of the Layers

My partner has indicated to you that the numbers and organization of what follows is not linear. This means that I'm going to assign the energies of the DNA layers to your existing numbers one through 12. They are not in a list, or in a row or organized according to importance or power. They are given to match the numbers in your singular, linear reality of how you must communicate, and lists are always linear for you.

However, you will also see that there is a grouping sequence I have chosen here, which will further help you to understand how even the 12 work together. So see the 12 in four groups of three layers each. The groups will have names also, and I will give you these and summarize them at the end.

Chapter Seven

DNA GROUP ONE
Layers One, Two, and Three

Kryon

Book 12

Chapter Seven
DNA GROUP ONE
Layers One, Two, and Three
by Kryon

DNA GROUP ONE: The Grounding Layers

The grounding layers are layers one, two and three. These refer to the closest energies to your dimensionality, and the easiest to understand for you. As we progress, the layers become more multidimensional and harder to comprehend. So the "grounding" is all about creating a foundation for the "house of understanding" as we discuss and build your understanding from the ground up. The grounding layers are, therefore, the foundation. The numerological implications are here, too, for the 1, 2, 3 add up to six, which is the Higher-Self layer energy (as you will see).

LAYER ONE: The Biological Layer (see page 322)

There is no "most important" DNA layer. But this one is the closest one to you and the "messenger" for all the others. It is the only one that resides firmly in your 3D world, yet it is as multidimensional as any of the others. Yet this is the one you can see!

Three billion chemical parts work together to create a reality that simply cannot be put into a logical, linear box, and those who peer at the chemistry in 3D will always have a puzzle. For this layer is the one that "reacts" to the other multidimensional layers. The puzzle for you reading this book is this: If the multidimensional layers reside in the 90 percent area of this 3D DNA chemistry, then are they part of Layer One or not? Are they apart, or are they not? Again, you are now linearizing the issue, and it can't be solved this

Chapter Seven

way. So the answer is that they are all together all the time in "the soup." [Remember?] So at this very early juncture in the teaching, you just have to understand the energies around them and not try to figure out "where" they are.

When you listen to a radio broadcast, do you have to know where it is being broadcast from? In fact, that answer is that *it is coming from your radio,* but it's being broadcast from another place (a radio tower somewhere or a studio somewhere). That still doesn't answer the question about "Where is the radio broadcast originating?" What if it's a simulcast of many announcers from many different areas? So where is it coming from now? The tower? The studio? Each location? The speaker in the radio? The true answer is you don't care; you just enjoy the program. This teaching is like that. Don't dwell on the linear puzzle, or you just won't understand what we are trying to give you. Relax and enjoy the program!

DNA is, therefore, the engine and the map within the example I had my partner give you. It is reactive and creative. It pushes and it pulls, and like the other multidimensional energies in the Universe, it is "aware" and biased to help you find the creator inside.

It is the "antenna" of body orchestration, and takes that which is multidimensional and converts it into information, then action. It is never alone, but it needs the "system" around it to allow it to work. Alone it does nothing, but with the rest of its entourage, it is "whole."

In 3D, it is seen as the Human genome, all of the parts of the double helix that make up the DNA molecule. Science sees it as complete and also a giant puzzle. For they look at it with their own bias, never fully ready to "see" what it might really be—the transmitter and receiver of all that is about you, including all your past Akashic lives, your remembrances, spiritual learning of the past, even your potential Lemurian roots.

Layer One is the layer that facilitates the communication with all the other layers. This "confluence of synchronicity" is the great unknown and unrecognized attribute of DNA itself, which all "thinks together" and works together as one. Science hasn't even seen this yet, but knows it must exist. It creates the bridge of a sentient, aware force within the body. It silently waits for signals from both Human consciousness and its own multidimensional memory. Together, they orchestrate the changes that are possible, which are "directed" by Human intent.

Three percent of it is the engine of biology, and the rest is the antenna of the multidimensional "soup" that directs the engine into action. It is this "multidimensional soup" that is the energy that the masters touched to create miracles in the Human body, for their miraculous touch only gave instructions to the DNA to change the 3D chemistry, something that any Human can now do alone. Such is the power of this layer, the only one that can be examined and looked at from your 3D lens.

KETER ETZ CHAYIM is the Hebrew *name of God* selected for it. I identified its name to my partner in 2003. My meaning for this Hebrew name of God, is "The Tree of Life." It represents the DNA chemical structure tree, and the building block of all life on Earth. But the Human is the only entity of divinity that can change its structure through his own will. Animals have DNA, too, and their Akash is represented also within their DNA, but for different reasons than for the Human. Even vegetables have it! [Kryon smile] But their purpose is to interface with the other life forms in a system that supports humanity—the only reason the vegetable evolved to the level it did.

Humans love puzzles. Why does the onion DNA have more genes than the Human being? What does it tell you? Perhaps the argument about genetic complexity being related to the number of genes in DNA is not accurate? Indeed! This should tell you that something more is involved in why the Human Being is seen as the top of the evolutionary ladder, but with fewer genes than an onion. Could it be that something else was added to make the Human genes far more complex than any others on the planet—so then it's not about how many, but about a quantum energy within them? More about that later as we study the layers.

Notice that each layer of DNA is an ancient name of God. Why would this be? Because the absolute truth, dear one, is that you are a piece of the creator—not creation, but the source energy of God. This means that I know you and you know me. It means that you always have been and always will be. This is the hardest thing for a Human to understand, and the reality of it is truly hidden in the 3D hologram you are all in. If it were actually known or seen, then it would take the test out of the earth's purpose to eventually create the beginning vibration of a new Universe. Therefore, it's about the celebration of who you are as my spiritual family.

We have chosen to give you this layer as *number one*. In a quantum state, no number is alone, for the *one* speaks to you right away, saying, *"I'm a number, and numbers are designed to be with numbers."* After all, no one created a system of numbers with only one number! Therefore, a single number, you might say, is incomplete without another number to make it "whole." Is this too esoteric for you? What did you expect from Kryon? [Smile]

So the *one* sits there missing the rest of the numbers. But it also "knows" that in the circle it sits within, it represents the first of a linear chain. It also knows that the last of the chain, the nine, is next to it [in the circle]! It also realizes that as a *one*, it is the center of everything, the beginning, the first number to have substance, leading to the other ones with a more complex makeup. All this is to say that the *number one* is almost like the center number. In a digital domain, it represents "on," where without it there would only be "off" or nothing at all. So it is more than just the first item. It is the center item and, therefore, responsible for all the others in the chain, since they are all derivatives of it (based on it and multiplications of it).

You have established the meaning of "New Beginnings" for the one. However, when you think of the energy of this 3D strand of DNA, it represents the center of all of DNA and, therefore, the key to communication. So don't let the significance of this being the first layer be diminished. It's not just that something has to start— far from it. For this layer is the core, the kingpin for all of them to work. Remember also that there is no such thing as a beginning in a quantum state, since there is no timeline.

Let's examine the complexity even more for a moment, for within the 3D chemistry hides a phenomenon that is not present in any other part of the Human body. This double helix contains factors that we don't discuss much, but that all play into it being one of the most profound ones. When you use intent, this layer "sees" your conscious thought and gives instructions to the protein-encoded portions (the 3 percent engine that controls your genes). When you give intent to wake up your Akashic experiences, this is the layer that must weigh the dynamics of a huge multidimensional storage system [your Akashic record] against the attributes of 3D [regular chemistry] within the Human body.

The most amazing thing about the double helix is the confluence of "knowing" all at once for more than 100 trillion other molecules of DNA! If you could see the inner workings of this body energy, it would not make sense. How can 100 trillion items all be "informed" at the same time? There is no chemical that races between the molecules, giving them the same message. There is no electric synapse that races around the body touching each of them with a spark of data. The answer is within the mystery of DNA, within the spirals and twists themselves, for each of the 100 trillion molecules has its own magnetic field! It's tiny, but the properties of magnetics are there, and each one has a field that overlaps the next one. The twists are a product of these magnetics and the symmetry is quite revealing if you start looking at DNA with a magnetic pattern in mind. That, of course, will eventually be discovered.

Those who study electronic pathways and attributes all know the mysterious ways of magnetic fields. In physics, magnetics is seen as a multidimensional energy. It is! This should give you a clue what is taking place, for here is the magnetic messenger of DNA—the engine of communication between DNA and all cellular structure. There is so much here that has not been discovered or even discussed.

Through a process called "induction," all the DNA changes together. If you had a microscopic multidimensional measuring device that could somehow measure what the DNA was "doing," you would be amazed at the synchronization of it. For DNA in the big toe is just as informed as the DNA in the brain. All 100 trillion molecules work together, since they all have the same exact *knowing*. DNA magnetic fields overlap. They overlap, since the physical DNA doesn't just "sit there" all pristine like a bunch of ladders ready to be observed in a row. The double helix is the form of DNA, but only seen that way when it's unraveled under a microscope. Native DNA—that is, the way it "lives"—is in clumps with other DNA. This insures that the magnetic fields overlap. They intertwine themselves all over the body, and this is the way DNA exists in cellular structure.

There has been speculation over the years by doctors and scientists, that the Human body must somehow have a "second brain." This is because of the things that science sees happen to the body. A spinal cord is severed, yet the signals somehow keep being sent to the heart and organs to keep functioning. DNA is the unsuspected answer, for you might say that this "confluence of DNA synchronizing" is, indeed, the missing pathway of body communication that is not acknowledged by science. Magnetic induction is "transmission of signals without wires, when one magnetic field overlaps another." Your body is a giant transformer filled with wireless information transmissions via the pathway of DNA confluence.

When you have more than 100 trillion molecules all "knowing" the same thing in tandem with each other, something else happens; it helps in the creation of the Human Merkabah—not the entire creation of it, but the part that represents the Human spirit. [Other parts are made from the *guide-set* you carry with you for life.] The Merkabah is still another Hebrew word that means "to ride." Your entire being "rides" on your Merkabah and extends out past your actual body size for approximately eight meters. Recognized as "the ascension vehicle" by many, the Merkabah is like your spiritual footprint on the planet and, indeed, carries within it all the attributes of your Akashic record. How do you think that information gets onto the Merkabah? Through the DNA! Later, we will show you that your entire personal earth Human history is carried in each molecule of DNA.

So this physical double helix has many attributes that hide from science. First, it's mainly made up of chemistry that is random, and that is the "multidimensional map" that talks to the protein-encoded parts of DNA. This is more than 90 percent of the genome, and often considered junk. But the very size and proportion of "who" is doing the work is very revealing, for the chemistry of the codes [gene-producing parts of DNA] are simply slaves to the 90 percent that are doing the "talking."

Second, DNA is the master synchronizer of the Human body, not the brain.

Third, the DNA family of more than 100 trillion parts creates a singularity. It has "awareness" of what you are doing, and your spiritual intent. It often follows karma, for that is the spiritual blueprint that sets up the genes. Of course, you knew that. Let me ask you a question: With that many "working parts," what would happen if that "aware and listening" system had no direction? The answer is "anything it wants." Without your conscious direction given to your body often, your cellular structure [DNA] has no boss, no instructions other than those you were born with.

Now you know why we ask you to "talk to the cells," for the great "listener" is the DNA! This is how consciousness can heal, can change chemistry, and can create more immune cells just through intent. Science has shown this is possible, and now perhaps it all starts to come together for you? You must be "the boss" to your body, and the DNA has a process whereby it is ready to listen to you.

SUMMARY: Layer One is the double helix and represents the chemistry you can see. This is the bridge layer, the one that contains the 3D parts and the multidimensional parts. It receives and transmits, for the job of Layer One's energy is to take information from the multidimensional layers and implement it upon your gene structure. Therefore, you might say that it really is central to how everything else works.

LAYER TWO: Your Life Lesson (see page 323)

Dear ones, who are you? Why are you here? Is there something you are supposed to accomplish or complete, or an attribute of yourself that you can discover and change? These questions are all wrapped up in what you may call your *life lesson*. If this layer could have a partner, it would be Layer Eight.

Now, in this linear lesson, we have to give you the energies one at a time, so a "partnership with Layer Eight" makes no sense, since we haven't gotten there yet. So we will have to step out of linearity here and actually give you a preview of Layer Eight in a moment.

Many would ask, *"Kryon, what is the difference between the energy karma gives us and our life lesson? Or are they the same?"*

Since everything is related to everything else, you might say that you can't take this completely out of the karmic energy bag; however, think of karma as a "push to place you in a groove of appropriateness." Karma is about the group around you, your place in it and the emotional energies that are developed from that. You might say that karma is the default guide of direction for all Humans on the planet. It creates four and five generations of firemen and policemen, doctors or military men. You "fall into the groove" of what you think you are supposed to do, or what your parents tell you, and do it!

There's more to the story of how this happens than the "my dad did it so I'll do it" argument. There is a propensity to go a certain direction and fulfill certain goals that are built into your DNA, that come in with your karma. It also represents unfinished business, feelings of passion about certain things you wish to do and, or course, your relationships with family and others. Some are actually driven by karma to find abusive partners! The reason is not to be punished, but so that they can work it out or find a way to solve the puzzle. Solutions of such things have always been the "energy generator" of humanity for the planet, and those solutions raise the vibration of the whole.

In the past, karma has driven all of the passions of Human Beings to *go and be and do*. It was seen, identified, and explained early by some of the core religions of the planet. It is only in recent times that it was dropped in favor of a reward and punishment system of spirituality. A karmic system requires you to look at the whys of things and try to work on them. A reward and punishment system allows total control of the leadership of the organization, who then give the members the *way to God* through rules developed by men. The masses love it due to its simplicity and, indeed, it has brought many to search for the creator, and it is appropriate. The truth is that anything that Humans develop to find their way to God in any culture is seen by God as an appropriate path, for the love of God does not care how you create a path to home.

In 1987, things began to shift on this planet, and as we have described many times in past communications, your actual reality began to change. New spiritual tools have become available, new processes and a new awareness are in play. An illumination shift is what we have called it and, indeed, its "nickname" has become the 11:11. It's almost a graduation of the old energy into the new, but at the same time many are realizing that it simply is a remembrance of what you once had.

With this new energy and new tools came an entirely new set of enablements for humanity. One of them was discussed with you in 1989 when I arrived, for in the first book of Kryon we told you that you could now drop the karmic attributes you came with and create your own "groove," and it would not be the one supplied to you by the system of time [karma]. We encouraged you to do this, since karma is a very long-term and slow process of learning over lifetimes. In this new energy, we asked you to take control of your life and decide for yourself what your passion would be.

Now, after all these years, many have done this, including my partner. With it sometimes comes a totally new set of friends and the departure of others who simply will not understand it when you disengage the game of karma or reject the current spiritual thinking. No longer interested in the family drama? You must be sick! [They say.] When you disengaged, many of the family walked away. The karmic rules of families are set and very difficult to see when you are immersed in them. The family member who is the "sand in your oyster" was designed that way and it's no accident that you have been dancing the drama dance all this time. Now you elect to step away, and the family member is furious that the dance partner won't dance anymore. They leave.

Karma is that way. It's ancient and slow. The new Human has choice, but with consequences. Now the responsibility issue must be understood and developed. No more victims, no more accidents, and total responsibility for all that is around them; these are the new understandings.

Life lessons, on the other hand, are far more personal. Although this attribute may, indeed, be related to a karmic energy, it stays with you when your karmic attribute is dropped. So you know it's more profound than karma and belongs to your individual core soul instead of a group. It is carried from life to life, just like karma, but in a different way. So it has "carry over" attributes just like karma. Karma is about situations with other Humans, unfinished business, feelings to be completed and a system of interaction. Life lessons are totally and completely personal, and only about you with you. Typical life lessons to be learned are listed here for you to see. Know that these only belong to you if you feel they do. Each Human has one or more, and each Human comes in with this as an overlay.

Chapter Seven

Once the Human solves them, this solution is carried to the next life and never has to be learned again.

Learn to love
Learn to listen
Learn to receive
Learn to love yourself
Learn to speak your truth
Learn how not to be a victim
Learn not to let anyone define you
Learn how to feel your own mastery
Learn how to live with other Humans
Learn how to get out of blaming others
Learn to move out of duality [drop your karma]
Learn to take care of yourself more than others
Learn that you deserve to be here, and are not dirty when you are born

Each one of these are profoundly personal. They don't involve family, a karmic groove or group energy. They are yours personally and you work on them for lifetimes, just like you do karma. However, in this new energy, they are all on the table to be solved. The *life lesson* issue is one that did not separate itself from karma until recently. Until 1987, the old energy hid any kind of understanding that you could actually change what seemed to be "the way you were." If you came into the planet as a fool, never listening to anyone, unhappy and seemingly unable to function with others, you were labeled and seen as difficult. That was it. The idea that you might be able to somehow "change your stripes" was not normally thought possible. New Age practitioners before this time had a tough job, since only a rare few could actually understand and be able to work with such a concept of total self-change.

TORAH ESER SPHIROT is the Hebrew *name of God* assigned to this layer. I identified this layer's name to my partner in 2003. My interpretation of it for this book is "the divine blueprint of law." Some may ask: Why that interpretation? Herein lies the complexity that some are not ready to hear, read about, or understand. The very word *law* sounds like men's rules. So I offer you this: What about the laws of nature, or the law of supply and demand, or the law of gravity? These are not written laws but expressions of "the way things work." You might then reinterpret this phrase to mean "the divine blueprint of the way things work" if you wish.

There is, indeed, a plan for the way things work within Humans. Some call it Human Nature, and many feel that this law creates people who are a certain way, and it's all part of the variety of life. Some are this way and others are that way, and the psychologists put them in boxes. There is an underlying feeling that they are this way for life, and that nothing changes them, for that's just "who they are." This is where the really good news is, that Humans can change any attribute about themselves they wish to. Nothing is cemented in place! Phobias, fears, personality type, talents—all are changeable in ways we have discussed before. The source for these gifts is in Layer Eight! (We told you we would get there). Layer Eight, briefly, is going to be discussed later as one of the two layers that carries the Akashic record of each Human Being. Its name and attributes will be discussed, but you already know a lot about it, since you know the number of it.

Let's discuss the numbers. This is Layer *Two*. Two is all about polarity, or the duality of life. Duality is defined as the clash of two concepts within a singular consciousness. Duality is, therefore, the carnal and biological Human Being who also is a piece of God. In Human history, it often represents the clash of good vs. evil, integrity vs. greed, and God and the Devil. It is so powerful a force that the basis of much of the world's religious mythology has a "fallen angel" who then is somehow responsible for all evil. You are then told that you walk through your life with both God and this fallen angel speaking into your ear of the things you are supposed to do. Isn't it interesting that even mythology starts with a sacredness? There is no darkness in the beginning. Evil is only created from light in this scenario.

The Human Being has full choice of whatever he wishes to do. He can be a Mother Teresa or an Adolf Hitler. The blackest black, the most evil thing on the planet, starts with the Human Being creating it from his own duality. He has free choice to go either way and there is no sacred energy on the planet that will keep him from it. This truly is the free choice that we speak of. The duality is different for many people. It sometimes represents a struggle with a "higher you" and a "lower you." Odd as this sounds, many have gone through this struggle and have realized that the "God part" of the Human is the familiar part, and that everything else is a trick.

This is, therefore, one of the most profound setups of the Human Being. Some say the test is about "finding the creator inside," and they would be correct, for indeed that is what this is about. Some would rather just say it's "finding the Higher-Self, or that part of you that is balanced and peaceful." Whatever you feel about it, Human Being, you are probably aware that it is different for each Human. Some seem at peace with life, and others constantly struggle to achieve it. This is a testament to the vast variety of Human paths, and also the fact that there simply is no one solution that works for everyone.

It is the bias of linearity that says that each Human is the same spiritually. Even metaphysical teachers often see a Human's spirituality just like the Human body, where every person has the same parts. They don't! Let's say there is someone in front of you with some issues they wish you to help them with. Imagine if you could read their Akashic record. How many lives have they had? What is in the "spiritual jar" of learning? Are they an infant or a master in spiritual experience and learning? Do they have a core karmic attribute or an advanced one? Are they an old soul or a new soul? Depending on these answers, you could have someone who is ready to receive the highest information that can be transmitted or one who needs to start at the beginning with concepts that some of you reading this book have know and been practicing for years.

So, do you see how the *two* and *duality* go together with the energy of *life lesson?* Truly, I tell you that when duality is in balance within the Human Being, you see the life lesson so much clearer! They, indeed, go together, and now you are just beginning to see how the numerology sometimes tells "the rest of the story" when it comes to solving the complex issues that you have come to solve. How could a Human even begin to discover and work on their life lessons without balance? So now you know another core bit of information; work on the duality issue first and many of the other things in the puzzle will start to come together. There is no Human who can begin the journey up the enlightenment road with a duality that is out of balance.

We could stop here, but we must mention again that these layers are all interactive and they work with each other, intermingling and mixing with one another. Just because we identify them separately doesn't mean they are apart from the others. My partner told you that, but here it is again. With each layer there comes other energies that they work with, and other layers that are often "partners" in the puzzle. In the case of Layer One, it is the hub, so it works with all the other 11. But Layer Two works especially well with Layer Eight.

Layer Eight is the full, personal Akashic record of the Human Being. Everything that they have ever been, learned, or have accomplished is here—every gender, every spiritual experience, every lifetime. This shakes hands with the search for duality and the life lesson of the Human, for it is this past energetic history that has created the attributes that are unique to the Human now. I just asked you a number of questions about who might be standing in front of you with issues to solve. Every Human has a setup with dozens of core issues that have come from earlier lives.

So, do you see how the puzzle is built? It really starts with the Akash, then spreads from there. When you are born, you receive all the Akashic information in every cell, in combination with the 3D inheritance factor—that is, your parent's participation—as well as what we call "the dust of the earth." This is our way of saying that everything that ever happened is actually part of the double helix, from the very beginning Human.

Eight in numerology is *responsibility and practical manifestation*. As we will see later, this is the perfect number to represent the Akashic record. It speaks of who you are and your responsibility to the earth, and your ability to structure yourself in 3D, even though you really are a multidimensional creature. It's actually quite interesting what happens when you are born, for you are no more quantum than at that moment. For the first few years of life, the 3D fights for control and usually wins. Ever see how small babies are enamored with things around them that you can't see? Those things are very real, but it takes awhile for the child to adjust to only seeing things in 3D. [Kryon smile]

SUMMARY - Layer Two

Layer Two is the divine blueprint of the law of duality. It represents a search for an individual's life lesson, and it works very well with the upcoming Layer Eight. See the "blueprint" as the interaction of this layer with the Akashic layer, for a blueprint is an English nickname for the "recorded plan." This, then, is the layer of DNA that is responsible for giving the Human Being a feeling of purpose and direction. It carries with it the attributes of life and all that has happened—not in archive form, but rather emotional form. Some of the layers carry the memories and some of the layers carry the tools for action from the memories. One layer may be the carpenter and one may be the wood, while another is the building plan. Even within two layers, do you begin to see how they work with each other?

LAYER THREE: Ascension and Activation (see page 324)

This may, indeed, be the most difficult layer to explain, my dear 3D Human family! So my partner, who knows where I'm going here, cautions me to be more practical in my explanation and carefully speak of what this layer is and is not.

"Kryon, we already don't understand! Why is there a DNA energy about ascension? That's a process, not something already in the body. And activation? Your partner already told us that we really don't activate DNA, but it activates itself. So what is this really all about?"

Let's back up for a moment and be less academic. Consider the overview for a moment—that the DNA of the Human body is starting to show what it really is. Each layer from now on is part of a master's tool box. Every master who walked the earth was biological. The gifts that some had are legendary—for instance, one touch could send healing energy to make the blind see and the crippled walk. Do you think the DNA of those masters was different, or just "activated"? The answer is that each of them had the same DNA you do. But all their layers were at 100%, where yours are barely at 30% (using 3D terms here). So you may start to get the picture that within you is the power to accomplish just about anything you have ever read about or heard about in scripture! You are not given new DNA as you go. But rather you are given the energy of enhancing what you already have.

Let's speak of the two words, and as we do, think *Master* and not *Human*. For here is something in the tool box that absolutely has to be there, since Humans have already shown you it's there through their miraculous powers! *Ascension* in this new energy is defined as "moving into the next lifetime without dying." It is, indeed, a process, but one that you already have the tools for. Is it really possible that a Human can do this? The answer is that thousands have already done it. They have voided their karma, balanced their duality, worked on their life lesson, and have literally become someone else! When you ask them, they will tell you that the "old" self is not even recognizable! This entire process they went through was directed by this layer of DNA.

Activation means "convert something into a reactive form." So this layer is truly one for mastery within you! You have several of these kinds of mastery layers, as you will see, but this is the ACTION layer. It is the one that directs the chemistry in your body (through the double helix) to create something that is not active into something that is. So you might say it is the "activator of chemistry in your body" to accomplish what your consciousness requires. I just gave you a hint of something that also needs to be discussed.

DNA doesn't sit there ready to be improved. Nothing is going to improve DNA! Instead, and healers know this, DNA sits there ready to be *directed*. Big difference. Assume there is a battery in the closet. You need it. There is an odd control on the outside panel of the battery that is a place for your hand, a depression to place your hand within. As you solve puzzles in life, you place your hand in the depression and the battery gives you more electricity! Somehow the battery is "sensing" what you have done, through your hand. Now, think of what is happening: Your hand is not controlling the battery's ability to give you electricity. Instead, your consciousness is, and your intents are being measured through your hand. The battery then responds and gives you what you need. So let me ask you some questions: Does the battery change? No. Are you "activating the battery"? Not really. It's already giving full electricity, but you're just not getting it all the time. So what are you doing? The answer is that you are enhancing the efficiency of the battery past the point it was a moment ago. The "activation" is the scenario of your intent, the measurement, and the battery giving more electricity.

The reason the word activation is in the name of this layer is because, like the battery, this layer is fully charged and ready to go, but is always measuring your spiritual vibration. As you solve the puzzles of duality, create peace in your life, and claim the power of the creator inside, DNA knows it and this layer is used to increase the power to your cellular structure to live longer, think in a balanced way, conquer fear, heal itself, and move into an entirely new paradigm. So this layer is *activating* YOU. It is activating you into the definition of ascension we gave.

So what does that tell you about those who are "activating DNA"? Again, as my partner discussed, they are blessed indeed and appropriate in the 3D construct, for it doesn't matter what you call something if it gets the job done. If you drive on a street that's sign is misspelled, does that mean you are not driving on it? No. It only means that there is a label change. It changes nothing about the purpose of the street. The one doing DNA activation is creating a process whereby your awareness is increased. This changes you and your vibration. It then changes the confluence vibration within the whole DNA molecule, the communication between the parts. This simply means that DNA is all "aware" of what you are doing spiritually, all the time. Remember, DNA is a quantum spiritual attribute of the Human body, and not all of it came from Earth (more on that later). So when this DNA healer helps you, he is really making your "measurement" different to the DNA, which then activates pieces and parts of your 3D chemistry that manifests the things you have been working on in 3D.

DNA is the antenna of the body, listening for the profundities of your epiphanies, the breakthroughs in the tight fabric of fear and frustration. It measures the heights of your joy, the peaks of your passion, and sees the smile on your face as you finally understand that the pathway to God has always been open and available to you. It responds by orchestrating cellular structure to enhance who you are, and to complement your life on Earth. It uses each layer of its own magnificence in the perfect ways that will create the soup of healing, a confluence of love creation, and an honoring of the Human Being's intent. Finally, Human Beings are beginning to understand the Mastery within!

NETZACH MERKAVA ELIYAHU is the Hebrew *name of God* we have chosen, and my interpretation for this book is "Ascension and Activation." These Hebrew names are powerful. They are not to be spoken or listened to in order to create power—not hardly. Instead, they are only guidelines for the honoring and celebration of the mastery layers they represent.

The number three is a perfect choice for this layer. Remember what *three* represents? It is the catalyst number, and I have already told you what that is. This layer never alters itself or changes in any way. Yet it is responsible for the potential transformation of your whole spiritual life. Three is a powerful number that's attributes are the attributes of the catalyst. In addition, the three is allied with joy and the inner child. Both of these are also attributes of discovery for the Human to work with Layer Three.

Joy is not overrated, for it is the glue of balance for a Human seeking a way out of the darkness of depression and suffering. A moment of joy can erase months of uncertainty and sorrow for a Human who is desperate to see the light of laughter. Joy is the staple of children, who seek it out, demand it, and use it to its fullest. The natural state of children is laughter and all that steps on it is inappropriate.

I have spoken many times of the energy of the Inner Child being that which belongs to all Humans, but something they eventually outgrow. Many have criticized Kryon for bringing in this energy, saying that they didn't have a good childhood, and that to actively try to remember it is to then relive horrible experiences. Those who have said this are so mired in their own childhood issues that they miss the entire point. They instead "bring this back to their own issue" instead of seeing the light of it, and their comments give this away in their lives, for everything they are continues to revolve around the victim they were as a child. Blessed is the one who learns to sever these things from their current lives, and instead build around what they know to be joyful in the *now*. The "inner child" is not *your* history, necessarily, but that of humanity in general. If you had a wonderful childhood, then use it to remember what it was like to have no problems, where the only thing you had to worry about was how long you could play. There were no money or relationship issues. Family was not a concern, and there were not many days of sadness for any reason, for you were always loved.

Chapter Seven

Kryon is not in a vacuum regarding the lives of those reading the words or listening to the channellings. We know that some of you didn't have this kind of childhood. So isn't it obvious that we don't wish you to revisit it? Instead, the idea here is to search for the inner child that *did* play—perhaps the one in make believe or when you were asleep. It represents the overall childlike attitude, joy, play, and peace. If you didn't have it, you still knew about it! You wanted it so badly that the fact you didn't get it wounded you! So now I ask you to create it as though it had happened. This not only helps you to relax in the now, but also helps to "heal" the Akash and bring you to a place of forgiveness to those who have seemed to have taken it away. A true inner child experience melts the heart and creates a light where no darkness or hatred can exist. No, Kryon knows very well who you are, dear Human Being. Create whatever child you wish to be, but then go there and wallow in the laughter and giggles of a space with no worry or drama. Feel your heart and life lighten as you do, for remember, there is always sunshine above the clouds, where the child is, and where you are free to romp in the grass with the creator—something you have done for eons.

This is the state of being that is next to mastery, for if you were to analyze the lives of the masters of Earth, they would all seem to be "childlike." This is because they had Layer Three working to its maximum. They were ascended beings, activating the entire potential of their biological structure. They were one with nature, and nature saw this and obeyed the rules they gave. Even when some of them were led away to be killed, they sang songs and smiled. Some would say they were crazy. But those who have been in that state know all about it. For in that state all that the Human can really see is the quantum view. The love of God is that way, for it allows peace where there shouldn't be peace, and solace when the 3D world says they should be shutting down.

Summary - Layer Three - Ascension and Activation

Layer Three is the Ascension and Activation layer. It never changes and can't be changed. Instead, it is what changes the other layers and the chemistry of the body to help with spiritual awakening. It is the "action" layer, since it is always "listening" for changes in our consciousness.

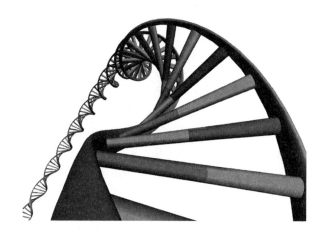

Chapter Eight

DNA GROUP TWO
Layers Four, Five, and Six

Kryon

Book 12

Chapter Eight
DNA GROUP TWO
Layers Four, Five, and Six
by Kryon

DNA GROUP TWO: The Human-Divine Layers

The Human divinity layers are Layers Four, Five and Six. As we progress, you will see that the energies represented in the DNA become more quantum [multidimensional], and less explainable. My partner knows this, and this is why he has turned all the explanations over to me. However, I remind him that he still stands as the interpreter within the channelling and, therefore, he is as much involved as I am in the process of converting nonlinear things to a linear thinking humanity. The numerological implications are here, too, for the four, five and six add up to six, just as the first grouping did. Layer Six is in this group, so you will see the profound meaning for it as we continue.

LAYERS FOUR AND FIVE: Your Angelic Name (see page 325)

Of all the layers, this one will be the one that creates those who will scratch their head in puzzlement. What is that about? And why two layers? It is the 3D thinker who will do this, for the only word we have in the language of Earth is "name." In the angelic realm, the identity of this energy is not singular. For you, it is. If I asked, "What is your name?," you would give me a word that you are identified by. You then join all the other "things" on the planet, like table, cloud, sky or rock. *"Wait a minute, Kryon,"* you might say. *"That's not true. For the rocks of the earth don't have separate names. We all have a unique name. So it isn't like a thing at all."* Fine, then let's name all the rocks and pebbles. They still are singular and don't carry an energy past 3D with the identification word.

We see you as a quantum soul group. It's time to again itemize the way this soul group splits when it comes into the planet. This is something we have channelled many times, but here it is again for this book, put into a condensed form.

First, I'm going to ask two questions that will help qualify what we are about to teach:

(1) *When you co-create for yourself, does it affect any other Human Being alive? When you actively create the thing you need, the home, the relationship, the job, the parking space* [Kryon smile], *are you doing it in a vacuum or does it affect any other Human Being?* Of course it does! Many! So here is the second question:

(2) *What gives you the right to steamroll your co-creation right over their lives?*

This would seem to be an unsolvable esoteric question of spiritual integrity, would it not? But when you realize the *system* of how everything works, then you begin to understand as best you can the appropriateness of it all.

During channelling with Kryon, the entourage that I bring often sits and washes your feet in love. You might wonder why we would do such a thing. What it is that is hiding so greatly that Spirit would come to you and kneel at your feet? Here is the answer, and it is one filled with awe and wonder in the profundity that it represents.

At the "wind of birth," as we call your birthday, we say goodbye to the part of you that belongs to the essence of God on my side of the veil, and an amazing series of multidimensional events takes place. Even the most elementary spiritual knowledge will tell you that you could not possibly exist on the earth in your full God form.

Therefore, you must intuitively know that something happens at birth to alter whatever you think your part of God is, to allow you to be born on the planet and exist here.

You split up into many parts, and some of them will surprise you, or you may actually disagree with what you are reading since your earth-based teaching is so singularly biased against what really takes place. The largest part of the split of your soul group is the Higher-Self. It separates from the Human Self in a quantum way. Now to a Human, the Higher-Self is "somewhere else"—perhaps in heaven. The truth is that the Higher-Self is really in the DNA! But it is not in 3D, and separates itself so completely from the Human 3D experience that it seems to the Human that it is not in the reality of your existence. This truth is common to most of the sacred and spiritual parts of the Human, that they are always with you, but seemingly not there since they are in a quantum state. Again, the Human sees things in a singular fashion and does not even like the concept of God as a group or themselves as a group. Everything in 3D is singular, with a singular time frame, life length, and with a nice start and finish. It's the "piece of string" syndrome, showing the incredible bias you have in 3D, and many can't think beyond it. You are just like the string—simple, singular, with a beginning and end.

The next thing to split out of your sacred soul group is what you call the guides and angels that are yours for life. This is controversial to a Human, since it voids all the nice illusions of what the guides are, their singular names, and what they do. The truth is that they are all pieces and parts of your own soul group. The concept of "you with you" is foreign to a Human Being, since it again is not singular and doesn't sit well with the logic of who you think you are.

"What?" some say. *"You mean my guides and angels are just me? What a disappointment."* This comment is prevalent to a Human at

the discovery of what we are saying, and it is said this way because most of humanity is told by spiritual leaders that they are born unworthy and that all good things come from above. The truth is that all good things, indeed, come from God, but that is within, and not somewhere else, resting on clouds playing harps.

There is a very interesting validation of what I'm telling you here about your guides and angels. Remember 21 years ago when I told you in Kryon Book One about the fact that when you change your vibration, your guide set retreats from you? It almost feels like you get new ones. Here is another piece of the puzzle that will fit for you, for the system allows for the Human Being to change the vibration or efficiency of his DNA. When that occurs, you have a personal epiphany, or realization or a "come to God" moment. It's called all manner of things, but it represents full spiritual life change, and an actual change in the way your own DNA communicates with you and the world around you. Suddenly, you believe fully in things that you didn't care about before or that were silly or not meaningful. The religions on the planet are filled with Humans doing this, for it is common to the spiritual attributes of humanity. It happens in every spiritual system, but is called different things. Still, look at the similarities and the issues around them.

When a Human makes this kind of DNA spiritual shift, the guides back away. They have to, for the Human is actually changing their vibration; their DNA is changing, and there is a period of time that is unsettling for them where the old things leave and the new things are seen more clearly. If you were to go from the mind of a child to the mind of an adult overnight, those around you would absolutely leave you alone, for only you can sort out the changes and start to be comfortable with the "new you." So it is the same with the guides and angels who are *you* and now back away while you readjust.

During this readjustment period, there is loneliness and a feeling of being left by a loved one. There are tears and almost a feeling of non-purpose. This is just your own DNA getting used to a new vibration, a new efficiency. It is common to man, and common to any spiritual shift. When balance has been achieved, the guides come back and feel almost like they are new! Why? Because they are now able to communicate far better than ever before. Even in spiritual systems that don't believe in guides or angels, the Human Being feels so much better about their own place in the Universe, on the earth and in their lives. But during the hours, days or months of readjustment, it feels so lonely!

There is a report that Christ was dying on the cross, hanging there. Indeed, he was dying, and if he had not been taken down early due to the "sunset rules of crucifixion" of the time in Jerusalem, he would have died right there and then. The report by the guards is that this "son of God" uttered the strangest words imaginable. *"My God, why have you forsaken me?"* This Jewish master was more in touch with God than any other Human at that time, and now he is suddenly alone? You see? He was going through an incredible shift that even separated the master from his very essence! This now explains this phrase to you, and shows again that the guides are so much a part of you that to have them back off even for a moment is traumatic. Little does the Human realize just how surrounded he is with the sacredness of the creator.

Even biblical historians argue over this phrase that the Christ used... saying it was purposeful "prose" of quoting past prophesy of his death, so that he would prove to all that he was who he was. Humans do many things, but it's a stretch even for the Christ to conjure up a poem or two as he hung in agony in his ninth hour.

So now we have two sets of the *split of you at birth*, on our way to the third one, and the answer to the co-creation integrity question asked originally. The final split is what we will call the "entourage of your soul group." This is not explainable to you in your logic and dimension bias, so just think of a bunch of *you* staying on the other side of the veil. Why? I'll explain in a moment.

With all of the essence and core of your spiritual soul being split out, you arrive as a baby Human Being on planet Earth, in 3D with a singular bias, and with only a fraction of who you were. All this you do willingly for the reasons we have given you before— caring about the Universe enough to help form what is to come. All this has been channelled before and is for another lesson. But know this: We see what has happened, and here you are reading this book! Do you know who you are? We do! So we see the incredible Human Being as a worker in the light, and one who has come here knowing that they won't remember anything until a day when their free choice touches them and asks to look. But notice what happens when the Human gets to a certain age. They look for the creator! Over 85% of humanity is involved in some kind of spiritual system, believing in the afterlife. This is a mandate of logic that just screams that something intuitive inside the cellular structure of humanity is missing something and goes looking to find it as soon as possible.

Again I say this: Blessed is the Human Being who finds the creator energy anywhere! Anywhere! For there is no judgment as to how they do it, with what doctrine or system. It may not be the one you have found, but if it soothes that part of them that needs to find solace and the love of God in their own way, then it is appropriate. More than appropriate, it is the first step to a journey of finding the Higher-Self in any way, with any words they choose. Healings occur in all the systems. Did you know that? Epiphanies and changed lives are there, too. Never let your singular bias get in the way of thinking that there is only one way to God. You are already God, and there are thousands of ways to "see God in you."

This last group [the entourage of you] stays on my side of the veil. They are, indeed, a part of your soul group. *"So,"* you ask, *"what do they do?"* This is such a linear question! It's very funny to Spirit. Yes, Spirit has fun with Human partners, and we smile and celebrate the learning that takes place with all of you who decide to enter this door of quantum lesson. They don't "do" anything, but they exist as a balance of humanity, providing for a bridge of your own mastery.

This entourage has no name or form. Think of it as "an energy representative" of your part of humanity, living on my side of the veil. Think of a congress of trillions of soul pieces who work together to support all of humanity, and especially the ones on the other side of the veil who have their essence [the Humans in 3D]. How does this work? Let me give the simplest explanation I can: They represent the potentials of every decision that might be made by the Human Being on Earth. They are swimming in quantumness, seemingly random potentials that only become structured with your permission and intent given on your side of the veil. We have told you that your intent represents one of the most powerful forces on the planet, for it changes your DNA, as well as provides for co-creation.

Ahh, co-creation! Here we are again at the beginning of this lesson when I asked the two questions. Now the answers: When you co-create, you do it in conjunction with this entourage. All the soul groups are involved who are around your co-creation. Did you notice how often your co-creation helps others? Even when you co-create a new job that puts you right where you want to be, did you notice that it's a win-win situation with the others around you? When you co-create a difficult situation where you may leave a relationship, did you notice that the one left behind eventually benefits from the lack of drama, or the life change that occurs because you did what you did? Now perhaps you are seeing the bigger picture? The entourage of *you* works with the entourage of all the other Humans around you to help facilitate the changes that are the best for all! So the answer is that there is more than just integrity in your co-creation, for it helps those around you.

Here, dear Human, is the secret behind how your co-creation for yourself helps all of humanity and the vibration of the earth. For what you do for yourself, seemingly just to move into appropriateness, drives the engine of vibrational change for the planet. Anything you are able to spiritually co-create helps everyone. Now you know. It's also why co-creation is timed the way it is, not in the time you expect! It's the biggest frustration to a Human Being to not know when these things are going to manifest. Quite often Humans feel that their manifestations must occur on their own schedule. They quickly learn that there is more to the puzzle, and now you know why, for the others around you are always involved in this sacred process. Think of yourselves, therefore, as the "leader manifestor," instead of just a co-creator of your own destiny, for it also helps those around you, with permission of their pieces of their own entourage, in a place you can't even imagine.

What's in a Name?

So, here comes Kryon into an event where Human Beings are seated [a live channelling]. Through my partner's reality, we sense each one in his chair, but we only feel the *angelic names* of who you are. This is a name that is an energy of recognition and respect for the piece of God you are in the Universe. Let me tell you what this "name" entails. First, we see all the parts that split out. We see the whole you, and this means we recognize a piece of us! For there is no singularity with God, no parts. The parts are only a product of what happens when you come into 3D. We are Multidimensional, and suddenly we "see" the quantumness of family within you. You don't see it, but we do, and often we are overwhelmed with joy at the reunion, for not often do Humans give intent for this—a process where they sit and let us walk among them and feel their souls as family with family.

Next, we "see" the whole Human experience, in that we are now looking at the DNA Akashic Record of each of you. So if you have had 100 lives on the planet, we see all of them. There is no *time* where we are, so we can actually relive every life you have within your own personal Akash represented in your DNA. We are, therefore, seeing your entire past-life spectrum of experience—everything you have ever done on Earth, ever learned, ever cried over, or celebrated. We are there at your birth and death, and all that happens in between your lives. We are there to see what you have learned that brings you to the chair, or your eyes to the page of this book.

So, my dear Human, all that I have said in the paragraphs above is *your name.* Do you see how large this is for us? It is also large within the context of how it must be represented in your DNA, for it requires a double energy when compared proportionately in 3D with the other layers or energies. Therefore, you might say that it is your God-family name, and it is huge!

This is the "name" represented in the Cave of Creation on your individual soul crystal, which we have discussed now for more than 20 years, for the Cave of Creation contains an interdimensional record of all the soul groups on the planet—one crystal for each one. Each crystal represents one Higher-Self, the same one each time. So even if you have had more than 100 lifetimes on the planet, it is still only represented by the one interdimensional crystal. It stays on the earth as long as the earth is here, and even if you are not. For the energy of what you and the others have done are part of the Gaia energy and remain. Therefore, you might say that your energy stamp of everything you have ever been remains on the earth. You activate the crystal each time you arrive and deactivate it each time you leave. But the energy of your historic significance remains like a library that is anchored to Gaia forever.

URIM VE TUMIM and **ALEPH ETZ ADONAI** are the two Hebrew *names of God* together, which we have assigned to your Angelic name layer group, DNA Layers Four and Five. The Kryon meaning for them is "The Light and the Power" and "The Core Crystal Energy." These meanings should be obvious within the description of "what's in a name" that you just had. You are, indeed, the light of the earth, and the only power that can make a difference. Your name, therefore, is a name of God itself, but one that is not in Hebrew, but rather one sung by us in light as we greet you. You are not in the reality of a piece of string, but rather as the fabric of time itself, always there without beginning or end, always with us in our wholeness of God. You are, therefore, core family!

The numbers are all very significant, for they speak alone, and in combination. Look at number *four*. The simple explanation for it is of Mother Earth, structure and understanding of structure in a physical world. This then is the singular part of you and your relationship to Gaia. Look next at the number *five,* for it represents pure change! This is the *Mars number* and can be a very fast energy—that is, one that creates the ability for the Human to change dimensionally. These numbers sitting together are linear, so they are comfortable with each other. The four tingles a little with the potential of the three, whose energy we discussed—very powerful with catalytic overlays. The five sits next to the grand one, the Higher-Self, so it is influenced to change toward the spiritual! This helps to explain why you are drawn to *find the creator*, as we have said before.

Now look at what the four and the five together add up to, for we are now presented with a *nine.* This is a psychic number and very sensitive. It's not just about completion, for that's too simple, but a completion of the meld between dimensionality, which your very angelic name implies within its existence in the DNA of a Human. So it's a sensitive completion that carries with it a sense of your ability to build a bridge to the other side, hence the psychic part of the number.

Summary - DNA Layers Four and Five

Layers four and five together are the essence of your expression (this specific life on Earth), and your divinity on the planet. They represent the "name" on the crystal on the Akashic Record. Together, they can be understood as: *The primary and most important spiritual attribute of all is the tree of life, which is family.* These are names of God and should never be thought of as separate layers. They are part of the divinity group It is so large in its energy that it takes up two layers or energy groups in the 12 that are given in this book.

LAYER SIX: The Higher-Self (see page 326)

Everything that has led up to this layer explanation has created a "bed of understanding" for this one. For now you know that this portion of the *God in you* splits out of the part of you that eventually becomes the Human Being. By now you know why it absolutely has to be in the DNA. The Higher-Self is the brother-sister that is absolute divinity. When you feel the creator inside, it is the Higher-Self that you are feeling. It *is* the connection to the other side.

Again, within humanity in general, there is the feeling that God is separate and that a Human is just biology. It is thought that through a series of epiphanies and decisions, a person can somehow join a larger energy and then be part of the God system. There is nothing odd about this, since it goes right along with the bias of singularity that is present in 3D. Humans believe in the singular, biological self. If they don't see something in 3D, therefore, it does not exist. Even with science telling them all about the invisible

Chapter Eight

dimensions around them, none of these things seem to apply to real life. Therefore, the Human is alone, helpless, and must somehow find an entirely different singular source for spiritual power. The search is complete when they find a spiritual system that suits their culture and perhaps what their peers have also found. Often, it changes their lives. But sometimes it is just an answer that makes them feel complete and no further search is needed. Then they give it honor, but don't change their lives. Their search is over.

To help balance this singular bias, the masters of the planet came to be, giving wonderful information about how the family works and what is possible. But they were seen as gods themselves, often killed, then worshiped in death, as we have already said. This is all very linear and totally understandable within the linear bias that is Human. Now, however, it's time to step away from that and see a more complete truth.

The result of this new thinking will create a God who is "far larger than the one you were originally told about." This will not make anything you have learned "wrong," but will enhance your understanding of the whole. A whole new relationship to Spirit emerges, and you move forward with an attitude that sees all the other spiritual systems as appropriate for those who use them. There is no judgment or singular bias about there only being "one way to God." It instead becomes a celebration of all the ways to God. For you will, indeed, meet all of the family eventually, and will realize that all of you are just "climbing the tree of truth" and finding the seeds one at a time and living through the growth of each lifetime as you climb the tree.

Therefore, the Higher-Self is not separate or apart from the Human Being. It is, indeed, *God in you*, and is the hardest thing to get used to. For again, Humans want to see God as something far more intelligent and lovable than part of themselves! The truth, however, is that this Higher-Self is very lovable, for it is part of the system of love itself and, indeed, is not singular. So it is always available in the DNA, and exists as part of the portal that has been described. This is why one of the most important life lessons is to learn to love yourself. This leads to the ability to interface with a Higher-Self that is part of you, and communicate with it.

If you are starting to think more interdimensional, then you can consider the DNA as a vehicle or a door to something else. Rather than having the DNA "contain" all this, you now are starting to realize that it only "gets you there." Does an elevator contain your destination? No. It's just a little box that takes you there. DNA has some of that same attribute as you become more multidimensional. Think of it as an elevator to the Higher-Self. For when you get "into it," you are transported to a place that is far larger than you could ever believe. Sometimes this is frustrating to a Human, for they can't contain all of the emotion or dimensional shift that often occurs when they begin the journey. This is complete with headaches, sleepless nights and even upset stomachs.

They look at all this and say, *"Why is Spirit making me sick, when all I want is to find the Higher-Self?"* The answer should be obvious, in that your search is shaking everything up. When you begin a journey of self-discovery, it involves all the cells, not just the brain. DNA is present in all of cellular structure, so everything is involved. We have also told you the secret to these issues is to "talk to your cells" and to Spirit and direct them: *"Only give me what I can take in a slow process."* This creates *you in charge* of even your own epiphany, and you receive it in your own comfort. Doesn't this also give you an idea of really who has the power in all this?

Humans will take advice from a channelling such as this, then using their linear thinking process, begin to shift, then stop and wonder if they are doing it right. *"What were the exact words again?*

Chapter Eight

Do I do it in the morning or the evening? How many times a day?" So here is a very truthful answer to these kinds of questions. Consider for a moment you are the king or queen of an entire empire. Everyone bows to you, and your subjects are lined up ready to receive the wisdom of the ages that you carry. You are totally in charge of everything, including when you get your meals, who you see, where you go, and when you teach your wisdom. Got this in your mind? Now, can you really ask a question, *"What exact words should I use to my cellular servants? Will they be there in the morning?"* The answer is that they don't care about protocol! They are hanging on your every word, and they don't sleep, just waiting for you to give them a direction—any direction! Do you see? Again, it's the linear bias that keeps a Human wondering if they are doing something right within a structure that they have created for themselves.

Human rules around spiritual doctrines are just that—Human rules. Step away from that and realize, *"If my Higher-Self is within me, then I have a direct line to divinity."* Indeed! That is a wonderful first step. Then the Higher-Self becomes more and more available to you due to your renewed attitude of the relationship you have to it. In some circles, it is called "The Holy Spirit." This is an energy that visits Human Beings when they need help to accomplish profound things. This is also the definition of the Higher-Self. So this is something that is acknowledged as a real part of universal spiritual energy, and it has many names.

Having this energy within your DNA is really like having a pathway to manifestation. For when you begin to get in touch with this part of you, things begin to change in your life. You start to realize more about the entirety of the energy around you and your divinity. Therefore, it might occur to you that this energy within you is the one that you wish to "tap" when you meditate or pray. Indeed, you would be correct. For in a Human's singularity, he wishes to pray "to someone or something more important than he is." The idea that you might be praying to yourself is totally unacceptable, so you look for an entity, even Kryon. Instead, we are telling you that the act of prayer and mediation should start as a celebration, then proceed to a manifestation of going through the portal, with the help of the Higher-Self into the place that allows you to have a relationship with the family that sits around you, the pieces and parts who split out from you. This is your connection to home, and it will feel that way. It's about communication with the whole set of *parts of you*. This is why we also call it the *Prayer and Communication* layer.

So we like to call the Higher-Self the prayer and communications layer, a pipeline or conduit to the other side of the veil. It uses your own sacred energy and this is important. Imagine this for a moment with me: The Human sits at the door to heaven. He wonders if he is worthy to open it, or even knock on it. Will he be turned away? He is just a Human. So he prostrates himself so that whoever opens the door will see the honor and take pity. Like this example? It represents most of humanity and the way they pray. Now, let's do it differently.

You are a member of the family and the door to your home stands before you. You can't go through it right now, but you can open it and communicate. On the other side are relatives and friends who have been gone for a long time. They don't know you are there standing at the door until you knock. Should you? You see? Now, as a family member, you have the right. Not only that, can't you see the joy they will have at knowing who is at the door? A reunion will ensue and the communications will be unforgettable, filled with tears of joy. Now, family member, do you prostrate yourself first? [Kryon smile] No. Spirit sees you as complete and worthy and equal to any entity in the Universe. The Higher-Self then becomes the vehicle you ride in to get there; the mastery rides upon this vehicle. More on that when we get to Layer Nine.

Chapter Eight

EHYEH ASHER EHYEH is the Hebrew *name of God* we have selected for this. This particular name is grand, indeed, for the meaning of it is "I am that I am." This phrase many of you have heard before. It is sacred in many spiritual systems, and the story is that this is who God said he was when Moses asked, *"Who is there?"* What can it really mean? It seems to be a puzzle. Actually, it's not. Only in languages other than Hebrew does it hold a funny syntax. The true meaning is, "I shall be what I shall be." This is a statement that still holds a wonder about interpretation. So think of it this way: "I AM" meaning this is my name; "I SHALL BE" meaning that I exist forever. Then you have a full sentence, actually. **"I am God, and shall be forever!"** This is, indeed, the energy of the Higher-Self, that of the creator, for it's the God part of each and every Human Being. It is the portal for communication to God.

Why is it "higher," some may ask. How do you see it, Human Being? Is it high in the clouds? Is it in heaven? We have already told you it's inside, so what is the meaning of *higher?* Now we also reveal that it is the part of you that *vibrates higher* than any of your 3D Human cellular structure. It is not "above" you, but rather it is vibrating at the frequency of the creator, and invisible and shielded. No wonder we ask you to open the door carefully!

The numerology of this layer is multi-layered in itself. When the *six* stands alone, we have given it the energy of communications, balance and harmony. This itself is a very nice compliment to the Higher-Self definition, but there is more. The *six* is next to the *five*, which is *change*, and also the *seven*, which means *sacred*. Do you see how this quantum numerology works? For you, indeed, have to consider where the number came from and what has been next to it in the linear or non-linear sequence it represents.

Now I will reveal that some of the layers are not complete unless they are working with some of the other layers. I told you that this would become more complex, and now it does. For the Higher-Self layer cannot stand by itself within the DNA system. The system depends upon interactivity with some of the specific energies around it, and some of the numbers in the *circle of 12*. Again, we call upon what my partner has already given you in the example of the automatic transmission. Even the main wheel that engages the engine must work with the smaller parts to determine the ratios and speeds it will attain. We encouraged you earlier to think of DNA in a circle of numbers instead of a list or stack. Think of it, then, as part of a DNA engine.

Layer Six works best with the energy of the *three*. Go back and read what has been said about DNA Layer Three and you may very well start to get the implications as to why this would be so. For the catalyst represents what is possible when one energy helps to create another while not changing itself. This is the explanation of the Human Being finding the Higher-Self in every day life. The Human changes greatly, and the Higher-Self remains the same.

Again, take a look at the *six* and the *three* working together, for they produce another *nine!* This is a common theme in DNA energies that work with one another, for completion in this situation is defined as the completion of opening the portal to the Higher-Self through the action of the catalyst of joy.

SUMMARY: DNA Layer Six - The Higher-Self

This layer is really a portal to the largest sacred part of you that split off when you came to the planet. It facilitates the very communication with God through prayer and meditation and is the "I AM" layer. It is seen as the pipeline to the other side of the veil, and is often called "The Holy Spirit." It is also the last layer in the Human divinity layer group and works with your multidimensional Layer Three.

Chapter Nine

DNA GROUP THREE
Layers Seven, Eight, and Nine

Kryon
Book 12

Chapter Nine
DNA GROUP THREE
Layers Seven, Eight, and Nine
by Kryon

DNA GROUP THREE: The Lemurian Layers

The Lemurian layers are Layers Seven, Eight and Nine. Again, as we continue, you will see that the energies represented in the DNA become even more interdimensional. In the case of these layers, however, they also tell a story of creation and the profundity of what actually might be in the DNA that makes it "the library of humanity on Earth."

LAYER SEVEN: Lemurian Layer One (see page 327)

Dear ones, how is it that we can ever explain something so odd and so strange to you? All of your major spiritual systems have the beginning of Human spiritual thinking coming from an angelic source. The metaphors are numerous, but there are very few that would tell you that your spiritual essence has come from the stars.

Notice first that all of the "beginning of spiritual awareness" stories happen to Humans who are just like you and have developed high consciousness. No spiritual story in any religious system starts with a Human in a cave, hunched over beating its recent kill with the bones of another animal. Instead, all of the metaphors give credence to an "advanced Human" who is at an evolutionary stage that seems to be very much like your own. This is important, for it is very accurate and true. Humans received the "seed of spiritual knowledge" when they had progressed to the point where the intellect could take it all in.

Chapter Nine

Even with this, however, the actual seeding of the Human race by the Pleiadian brothers and sisters took thousands of years. It was not a quick happening, and required generation after generation of birth and death of humanity to accomplish it.

As my partner has described earlier, humanity had developed along the lines of almost all the other mammals on the planet, which created tremendous variety for survival. If you look around, this is the way nature works, for it creates variations on a theme, insuring the survival of the species. However, in order for this planet to have a spiritual consciousness and become "the only planet of free choice," there could only be one type of Human Being that would survive.

The "only planet of free choice" is a puzzling concept to many. Let me tell you again that this means that currently in the Universe, the earth is the only place where a spiritual being has full choice to accept or reject the creator inside. It's the only planet that can control the vibration of the Gaia energy or vibratory level of the earth. Therefore, it can "choose" how high to vibrate in a system that will eventually help decide the beginning vibration of yet another Universe. No other planet has this system. Others used to, and one of them is still orbiting around one of the stars in the Seven Sisters system.

Universes are created all the time, and the folly of your "Big Bang" theory is that it is (1) the product of Human scientific 3D bias, which is always looking for the beginning of everything. However, anything with a quantum attribute has no beginning. It doesn't seem to bother science that the atom itself has no beginning, for it was involved in starting the bang, but the Universe itself *must* have one! So therefore, as science sees it, the Big Bang all starts with atoms that are somehow all God particles, in that they have no beginning or ending, creating an explosion that becomes the beginning of your Universe.

(2) In order for your Universe to have this false beginning, which fulfills the bias of linear Human thinking, the indestructible atom is somehow activated in a mysterious way. In a nano-moment (as goes the thinking), *all that is* appears from *nothing that is*. In that creative moment, all the rules of physics were suspended to allow for this beginning and the Universe began. All this is given to satisfy the "how did it begin" Human bias, like a straight line in your geometry, which must always begin somewhere, somehow. That string can go forever, on its way to ending somehow, but it's your 3D linear thinking that demands it have a beginning.

Let me pause for a moment to give credibility to time, for it indeed weighs heavily on the Human logic process. Science looks back in time through fine telescopes and can see a developing Universe [via the light from billions of years ago that shows this]. This just cements the idea that there was a beginning, a time when galaxies were not as they are today. Therefore, development scenarios seem to support the idea that there once was a beginning and galaxies formed from something that was without form and new.

Let me again tell you, since we are on a space subject, that your Universe always was. It always will be, but there is a dimensional cycle. The cycle does not have the Universe collapsing upon itself. Instead, it has the Universe growing to a place where it must divide due to the quantum rules of energy. When this happens, an interdimensional shift occurs, almost seeming like a collision of dimensions. Indeed, this is what it is, for the quantum rules that define dimensional attributes have energy as their qualifier, and even a quantum state changes from one kind to another. In 3D, this might seem to be an *explosion*.

When that occurs, what you see is that 3D matter reconfigures itself. Life as you know it ends and then recreates itself after the matter slowly reconstructs itself into a new dimensional state. In

Chapter Nine

other words, one Universe splits to include another, but it simply is a continuation of the other, but in a higher form. So in a way, there is indeed a *beginning* to your current configuration, but it's only a *continuation of what was.*

Hiding in all this are the influences of what you call the "black holes," which are really the dimensional "parents" of each galaxy. They provide a balanced push-pull energy that keeps the galaxy moving as one, and glues it together with quantum attributes you have yet to discover.

It is these "black holes" that are the secret to your dark matter and also to the energy quotient of the Universe, and why there seems to always be an expansion going on. It's not really a size expansion as much as a quantum energy expansion. A quantum energy expansion is an attribute of energy increasing in dimensionality, not in size. As it does, it displaces things that you perceive in 3D as distance. This perception is simply the fact that you don't yet see interdimensional processes. Are galaxies racing away from each other? In 3D, yes! However, in a multidimensional state, they are not. They are the same, but what you see as "dark matter" is growing.

The truth is that energy is being created that is invisible, and it pushes the matter in 3D away from itself. If you were a molecule inside a balloon being inflated, you would perceive only that the skin of the balloon was retreating from the middle. It was racing away from you at an ever-increasing speed. What you would not be aware of was that there was more invisible air being pumped in. So energy, which is "invisible" to you as an interdimensional force, is being created all the time, but you haven't even defined this kind of energy yet, and you don't even acknowledge it. You only know it's there in a tremendously large way, but don't know what it is. These are items you are only beginning to conceive, so it will take some time before there is an acknowledgement of the science behind all this. When that occurs, remember where you heard it! [Kryon wink]

The energy "residue" that science points to as proof of the Big Bang is only proof that something changed and a large shift occurred. That is the way of it, for a dimensional shift is what occurred, not a situation where everything magically came from nothing in an explosion that is unexplainable and whose atomic structure is a mystery. Do you see where Human bias sometimes takes precedent over logical science? You are biased to think in a straight line, and that line must have a beginning.

Enter the Pleiadian...

By design, a very advanced race of spiritual beings began to work with humanity. They had been on the earth many times, waiting for the right moment to begin their work. They remained, in their own way, and made certain that the DNA they were changing within humanity was correct.

They were here by divine plan. They did not wish to conquer or claim the earth. They did not wish to seed the earth with their kind, then return and take over. All these kinds of stories are Human created and they show a fear-based bias, which is very Human, not Pleiadian. Still, conspiratorial stories remain that say these beings are "hiding" within Humans and will someday, somehow capture your souls or have some kind of plan that is sinister.

So, dear Human, may I ask you: Do you see God as invasive in your life? Is God here to capture your soul and take you somewhere against your Human will? Does God seem to you to be the ultimate conqueror of life force? Do you have God walking the streets of the planet, spying on Humans so that "God Central" can be informed about what you are doing? The answer is obvious, but still there are Humans who want the drama and wish to have the conspiracy to make things exciting. This is their free choice. Is it yours?

The Pleiadians represent the graduate life forms of the Universe as you know it. They were the original "only planet of free choice" and had a spiritual influence on the energy of the very creation of your solar system. Their civilization is older than yours. Not by much in Universe terms, but enough so that they developed and went through their wars and tribulations and graduated with a task—to continue the work by giving their vibration to the next generation of Humanoid forms. They did this by passing on the quantum [interdimensional] portion of their DNA, right to the mammal called Human on the planet Earth. It was time. Earth was ready. It was approximately 100,000 Earth years ago—not too long ago really, compared to how long it took for Earth to get to the place where it was possible.

What we have not described to you was the fact that the Pleiadians were not the first off planet visitors. There is no reason to go into this, for it has nothing to do with your spirituality. Suffice it to say that by the time the Pleiadians began their work, the others were no longer interested in what Earth had to offer. There were no star wars, no battles to fight. They did their work at a perfect time, ordained by the creator energy to be correct and proper for the next evolutionary step of Earth—the seeding of the creator energy within Human DNA.

The Pleiadians have quantum [interdimensional] technology and they understand the additional two rules of physics that you have yet to discover. This gives them the ability to entangle themselves with the quantum state of the Universe and travel to you almost instantly. Knowing this, you can see why it's not needed for them to hide in Human form, deceive or trick, or disguise themselves. This is all very silly to them, for they are here any time they wish to be.

Today, Human technology regularly monitors areas with "long distance" sight via cameras that continually allow visual pictures of areas where Humans are not present. In other words, you can visually see anything where a camera is installed. Even as little as 100 years ago, this attribute would have created fear in a population that could see you do it. You would be labeled "evil" and witch-like, since you could invisibly see what was at a distance. It would lead to conspiracy theories and fear-based decisions. Humans back then would simply assume that you were "in the room with them" instead of just sitting watching a video screen somewhere else. Do you see? Your perception of reality is shaped by only what you know, and humanity has always created evil and fear out of what they do not know.

The Pleiadian interdimensional craft is also notable and has been seen. Things in an interdimensional state are simply not going to move in a 3D manner. When you hear of objects "shimmering" and moving at impossible speeds, only to stop instantly, then do it again, you are seeing movement in a dimensionality that is odd and not yours. This is not anything you have in your science yet, for everything you have still moves in one kind of dimensionality—yours. This limits everything to your own 3D expectations. However, science already has seen signs of matter in an entangled state—that is, being seemingly in two 3D places at once. Think of it this way: You are everywhere at once! Draw a circle in the sand; ants are in the circle. You place your finger in one place and startle an ant. Suddenly, you place your finger in another place almost instantly. The ant thinks you have moved from one place to another at frightening speed! All it sees is the finger. Understand? You are always above the circle, and you didn't "travel anywhere." This is entanglement, and it is a staple of interdimensional "travel."

The 2D stick figure cartoons on the flat piece of paper are astounded by the spooky visitations of the seemingly weird and unexplainable "lights on their page." They confound their logic and even startle and scare them. What could it be? Meanwhile, the 3D creatures hover over the paper, trying their best not to startle their friends, who simply don't have the concept of "up" yet.

Your Pleiadian brothers and sisters were not DNA alteration experts. They didn't reconstruct Human DNA or add to it. They did not construct laboratories and perform microsurgery on a few Humans, sending them out into Human culture. They did not abduct or frighten. Instead, they did what you think they did, and through a process of integration of birth attributes, slowly created a hybrid of themselves. The result was a Human Being with their spiritual DNA attributes—something that was completely missing on the earth. Also in the process, humanity became part Pleiadian. You have their DNA attributes, the spiritual ones that they gave you through the normal birth process of biological inheritance. Does it occur to you that they did this without startling anyone? Indeed, they did, for they "fit in" to the Human society. This also should tell you that they look just like you.

How would you do this if you were them? It was done with tremendous integrity and love. They lived with you. They spent years and years, even beyond their own life spans (which is considerable), continuing the process for dozens and dozens of Human generations. Then when it was finished, they left.

Let me again tell you something, my dear Human family. Someday when they return and provide a 3D meeting experience for you, when you see them, you will laugh! For they look just like you! A bit taller, perhaps, but they will shock you, for it will be so very obvious who they are—your star seed parents, and a loving group they remain.

Slowly, the quantum portions of your DNA received what the Pleiadians wanted you to have—a system they had that includes the creator energy, the Akashic Record and all the other attributes that you today call spiritual. So the three "Lemurian Layers" represent the core of what they gave you. So this layer is the "Extra Dimensional Sense Layer." Again, you can't really count what they gave you in layers or items for it is spread all over your quantum DNA. But in this explanation, I can give you the main elements in a linear fashion, and so I have selected these three energies as the Lemurian Layers.

Why Lemurian? Because as the other kinds of Humans died out on the planet, the ones with the Pleiadian DNA survived and began to create their own cultures. Although the Pleiadian seeds were given worldwide, there was one large civilization that gained incredible understanding and strength. Humans don't always evolve the same. Look around you today. Is everyone the same? Do those in Africa have what you in North America have in stable government, resources, invention, technology and wisdom of culture? No, they don't. Yet they have been around as long as you have. So you see, there is great variety in the ways Humans work with what they have, or don't.

As described so many times, the greatest society was in the middle of the current Pacific Ocean, which later became the Hawaiian Islands. Therefore, you might say that the Polynesians were the first kinds of Humans to gain the full knowledge of what the Pleiadians did, and you would be right. Note, for those interested, that the first kind of fully realized Spiritual Human was one of color. [Kryon wink]

Lemuria is, therefore, the first civilization of the planet with Humans like you. Oh, there were many others with the quantum DNA all over the earth, but none advanced like those on Lemuria. The language of Lemuria has been lost, but it survives in a modified core fashion with the "ancestral language of Hawaii," for this is their lineage, and that of those who left the *big mountain* due to rising water and fanned out in all directions to eventually seed the earth. They first went where the currents took them naturally, then later sailed against the wind.

Many have asked, *"Kryon, does the current Hawaiian language therefore sound like Lemurian?"* Somewhat, but not as much as you would think. As is the way of things, language morphs with time and much is added and subtracted. This is not unusual and the time needed for all this to happen was considerable (to a Human). So today you really have nothing that sounds like full Lemurian. If I had to give you the closest current languages, however, it would be a combination of Japanese and Hawaiian. As the Lemurian population scattered, it created many new cultures and eventually contributed to many languages of Earth.

"Kryon, science does not acknowledge Lemuria, and says that the seat of Human civilization came out of the Middle East. What do you say?" I say that the truth is the truth. As I described, the Pleiadian seed was planted on all continents. So it's what science has discovered that creates their facts. As discoveries shift, their facts shift. Science will continue to change with discovery. The truth stays the same—that in the middle of the north Pacific Ocean was the most advanced and oldest stable Human civilization in history.

Let's make something else clear. Before the Pleiadians did their work with you all over this planet, there was no divinity inside the Human structure. You had no creator inside, and the system of your divine coming and going did not exist. You were the same as much of the other life in the Universe, for there is an assumption by Humans that all life is the same. It isn't. There is only a fraction of life in the Universe that is "allied with the divinity of the creator," and you are one of them. Thanks to your brothers and sisters who helped put this in place, you are part of their lineage and part of their planet's purpose. By the way, this means that at some point, there was a "first"—the first Human who embodied the creator inside. It wasn't easy, and those around him were not yet ready. It took centuries to bring all humanity into focus and to give them the DNA we study today in these pages. In the process, there was difficulty. In the process, there was fear. Don't believe that this was done instantly or overnight. The first Human was killed by the others. They saw his light and they were afraid. Many others went through the same thing, bringing into focus what is today called, "fear of enlightenment." Old souls have it, for good reason.

If these things interest you further, we have given many channellings about Lemuria, the lifestyle and the time frames of a developing humanity. This is not the place to review them. But know that again I tell you that the process was blessed by God and very appropriate in the eyes of God. The Pleiadians were following the intuition and synchronicity of Spirit to come to Earth and seed one kind of Human with their spiritual quantum DNA attribute. You simply did not have it before they got here.

Because these energies (layers) seven and eight are labeled "Lemurian Layers One and Two," they are very special in the scheme of Human spiritual history. For this reason, I will actually give you not only the Hebrew names for them, but also the full Lemurian names and meanings. For to me, Lemurian is still a language.

KADUMAH ELOHIM is the Hebrew *name of God* associated with this layer. The Kryon meaning is, **DNA Home Language** and **Revealed Divinity.** What we are trying to say in this meaning is that the language of "home" is now inserted within the Human Being. "Home" is not Lemuria, nor is it within the Pleiadian system. Home is the side of the veil that I represent and from which all love is generated. "Home" is, therefore, the place you believe God resides. What we have called "The Third Language" is represented by this layer. Notice the three? This number is again the catalyst for change. So the "third language" is now revealed to be the "Lemurian connection" and "the quantum language of divine revelation."

So the very attribute of divine realization is within this layer. It represents the "search for the creator," which Humans intuitively have from the time they are born. They are looking for what is missing, and what fuels that search is this layer within them.

HOA YAWEE MARU is the Lemurian language name for this layer. I gave all this to my partner years ago in channel, and he has had all this time to understand the concepts so that when this book was written, he would have also been teaching it. It's that important that you understand the energy and the time that has gone into these explanations. Now, as I channel through him, these concepts are known by him and the explanations, therefore, are far less cryptic.

The Lemurian meaning of Hoa Yawee Maru is "revealed divinity." Without it, you have darkness, but with it, you have a reason to be here. It is the end of the age of innocence and so this is the equivalent of the realization of God. It is the exposure of the concepts of dark and light, of the consciousness of integrity vs. non-integrity, and of what others tell you is good and evil. However, you already know that we call it "that which reveals God in humanity."

The number *seven* represents divinity, wholeness, perfection and what we often call the number of "learning life." For the epiphanies of the Human are all connected with divine learning at every stage. Do you see how the number seven is, therefore, the perfect match for this first Lemurian layer? [Lemurian Layer One] In addition, we have the number one! Did you notice? It's part of the name, so you must consider how it affects the whole. So the one is "new beginnings" and also has been seen as "about self." Can you see how both these numbers tell a story about this layer?

If you knew nothing else but the numbers, you would know this: Here is an energy of divine wholeness. It represents perfection, a new start, and it is about self. Therefore, it has to be one of revelation for humanity. It is perhaps no better displayed than here, how the numbers work to give you a full picture of the energy behind the language you are reading that has linear words. [English] There is far more here to glean than just the words on the page, dear one. Can you feel the celebration we have regarding this layer and the fact that after all these years, you may now read about this? This is the miracle of the Great Shift and it speaks volumes to how far you have come—that even a book of this sort, with the truths presented here, could have ever been accepted.

SUMMARY: DNA Layer Seven - Lemurian Layer One

This is one of three Lemurian layers given to humanity almost 100,000 years ago by the Pleiadians. It eventually produced many cultures, but the most advanced was the civilization of Lemuria, which traces back to approximately 50,000 years ago. Note that the 50,000 years in between the seeding process and the beginning civilization represents the time it took to (1) create a new DNA type within humanity and (2) allow for the many other types of humans to die out, leaving only one basic kind of Human remaining on the planet without the kinds of variations you find in other mammals. It is a divine revelation layer, the metaphor of which is *the end of innocence and the beginning of spiritual awareness.*

LAYER EIGHT: The Master Akashic Record (see page 328)

This layer is *Lemurian Layer Two* and represents one of the most profound energies within Human DNA. It is a major part of the quantum portion of the double helix and represents the **Master Akashic Record**. I will also tell you that it works with DNA Layer Two, which you have already studied.

Chapter Nine

Let's review for a moment what this is all about. When we speak of the Akash, our definition becomes, "the energies of Gaia, as referenced to humanity." So it's actually a Gaia-based, or Earth-based, attribute. But it can't exist without the Human Being. Within the Human is the personal record of all that has happened to him or her.

In the beginning transcriptions of Kryon, we told you about the Cave of Creation. It's a real 3D cave, but also one with quantum attributes. This is not that unusual, for a Human Being is the same since it has both 3D and quantum attributes. This cave represents "humanity on Earth" and is crystalline in nature. It will never be found due to where it is, how deep it is, and the fact that it is hidden. No Human can enter it, for it does not support Human life as you know it. This is all appropriate, for it's not for Humans to discover. Perhaps you could call it "Earth-based spiritual accounting." The Cave of Creation contains a record of each and every soul that has ever been on the planet or is scheduled to be on the planet. This is the quantum portion, since there is no fortune telling here, but only a constantly changing dynamic of what humanity might do next regarding the constant coming and going to Earth.

Therefore, the Cave of Creation is a planetary record of all Humans and, therefore, is the Akashic Record of all humanity. Each soul has a crystalline record, and each lifetime is imbued into the soul's crystal. Therefore, you don't have a crystal for each Human life, but rather for each Human soul. Many lifetimes are then etched upon one crystalline structure, representing the one soul that has incarnated many times.

The DNA, unlike the cave, is personal. It carries the full Akashic Record of the one Human Being it represents. It represents the record of every lifetime that Human has had. It does not do it in 3D, so there are no layers of lifetimes or drawers with "year labels" on them, etc. Instead, the records are all in a quantum way, which is a mixture of everything that has been learned, seemingly in one energy.

This record within the DNA contains the interdimensional aspects of personal energy delivery, for it helps to posture who you are in each lifetime, apart from the biological, 3D chemistry contained in the protein-encoded portions of the DNA. This record is contained in the quantum, random chemistry and is not a linear representation of past lives. Instead, it is an "instruction set to connect to the main library," which is in a quantum state in what you would perceive as another dimension. I tell you this since there is an idea that the Human's Akashic Record is somehow physically contained in the DNA. It isn't; only "pointers to the library" are there. But even the pointers take up 3D space within the Human Genome.

I don't expect you to understand these things, but only to esoterically acknowledge them as the way it works. But it helps us to let you know that it's not linear, for it's much too large an energy to be contained in your DNA. Now I will also tell you that your angelic name, or energy as represented in Layers Four and Five, is also this way. We have waited to tell you this because as we progress to the higher layer energies, the puzzle becomes even more quantum, seemingly random, and harder to comprehend.

This Akashic Record contains not only past lives, but karma. It carries with it the reasoning for who your parents are, the karmic grouping you have chosen and the challenges you face. It creates the fears, propensities for drama, wisdom, peace and solutions. It is emotional, in that it "remembers" pieces and parts of past lives that create blocks and even habits. There is far more than meets a first look at what makes you who you are. As my partner has discussed before, the Human comes in with biologically inherited DNA traits from the chemistry in 3D of the parent, but the largest traits of the Human personality are carried by the Akashic Record,

Chapter Nine

and the experience that a Human has had on Earth in what you call the past.

This layer contains the setup of life and is also a challenging layer. This is why it works with Layer Two, which is the life lesson layer, and deals with duality. Do you see how these are related? No DNA energy layer stands alone.

Let's discuss for a moment the entire past life issue. Humans want to think of these lives as a collection of pieces of string (the Human bias again). They want to think each one is a specific length and each one is stored in a stack. They want to think the further away in Human history the past life, the harder it is to find or access, since it happened a long time ago.

It's not that way. Everything that has ever happened to you makes up the master painting of who you are today. As the cosmic artist mixes the paint and combines the colors, these are the lives that you have lived to help posture your energy and spirituality to the place it is today. Not only that, but Earth-based experience is also all there. The lives as a mother, a warrior, an accountant or a farmer are there. Therefore, you have experienced almost everything a Human can experience and it's all in the DNA represented, in this metaphor, by one grand painting that continues to expand and change.

Many wish to access a past life like they wish to access a layer of DNA. They want the simplicity of the 3D linear way, and it just isn't possible. Think for a moment: You are in a master art gallery and are sitting in front of a painting. Let's call it "George." You love it. There it is in all its glory and you are mesmerized by its life and beauty. You admire it.

Along comes a person who knows nothing about art and he can't see the painting for its entirety. Instead, he is analyzing the parts of it and saying odd things: *"I would like to see the pigment used in that red area of George,"* he says. *"I wonder if I can scrape away some paint and find that?"* You are amazed! Doesn't he know that this pigment was mixed carefully by the artist when he painted the canvas 100 years ago? Doesn't he know that the original color pigment is long gone, since it was then mixed with other pigments, then the finishing varnish over that? The answer is no, he does not.

Trying to capture the essence of a past life within the DNA is just like that. It isn't in a box or on a timeline in your DNA. Instead, it's part of the artwork of you. On the surface you can understand the energy of it, for there are those who can go in and linearize it for you and "read" a past life profile. However, this is not "accessing it" for the purpose of using it now. A past life reading is only sensing what it was and how it affects the entire painting.

You can read about a famous author or painter, but you can't bring them back from the grave and talk to them. This is because they are mixed into the fabric of Earth's energy and the 3D portion of their life has passed. But their energy remains on the planet, and so do their writings and art. Therefore, you might say they are here, manifest through the consciousness of their art. The quantum part is here, but the physical part is not.

Still, there are those who wish to work with a past life in 3D, as though it was still with them in a linear way. It isn't. Personally, it's simply part of you in a way that enhances who you are. It's part of the soup of your life essence and is one of the colors in the grand painting.

Mining the Akash

The entire reason we bring this up is that in this new energy we are able to tell you that you now have the ability to bring forward to your current life some of the attributes of a past life. Yet, we are not

Chapter Nine

saying that you can go identify a life and somehow use it. We are instead saying that your DNA "knows" what you have experienced, who you have been and how you were "painted."

The shift that is upon this planet is increasing your abilities, and your DNA is responding. Children are being born with more quantum (conceptual) awareness, and you are starting to realize more about the creator in you. The mastery that you have heard about that is inside you is beginning to be realized instead of something you just "heard about," and the invitation is here to start becoming more quantum and far less linear. In that process, there is what we then call "Mining the Akash."

All the lifetimes you have lived were stepping stones for the one you are now living. Each and every one gave you information, experience and wisdom for this one. *"What?"* some say, *"I know lots of people who are just not that aware and could care less about spiritual things. They have no wisdom at all, but a past life reader says they are an old soul! So what's that about, Kryon?"*

Here is where it gets complicated. Everything from now on requires a purposeful, realized intent. That means that although you might have the "library of experience of a master," it lays there until you actually believe you can use it. It can't simply be a curiosity, but a firm belief. You have to "own" the concept before it can happen, and many have no idea how to do that. They like the idea but simply cannot understand how to implement it. This is something we cannot teach, for you either are on the path of belief or you are not. Many give words that say they are, but their lives show they are not. Often it's due to fear, but most of the time they simply are not ready.

For those who are ready, the process is a magnificent one. Since it isn't linear, the Human can't say, *"I would like to activate lifetime 102, where I had the skills of an orator and was an author."* Even if that person is certain it was life 102, because they had a reading, that attribute of history no longer exists. Instead, the energy of it exists in the soup of who you are today. It hides, ready to be brought out and used again. It's not the "you" of today, but of one you already lived and experienced, so it is indeed part of you. Therefore, the process is quantum and you might address it like this: *"Dear cellular structure, I wish to have the attributes that I have earned in what I call my past that will enhance the ability for me to live my current life with more ease and grace. I wish to recall those things that will allow me to live longer, do my work better, and give me peace over the things that I desire to do. I wish to mine the Akash for this, which is in my personal DNA."*

So, do you see what is happening here? You are giving directions to your cellular structure, but not specifics. You are letting your own DNA "decide" for you what lifetimes might be involved. Do you understand the profundity of this?

When you accelerate the vehicle, you give instructions to the very complex automatic transmission to shift through the gears and move forward. You do not consciously address each gear, telling it what to do and how fast to rotate, what formula to use in fluid dynamics, or how to know when to interface with the gear next to it. The automatic transmission "knows" what to do when you tell it to "go." The DNA is like this, for it responds to your overall intent.

You can also directly address your cellular structure (DNA) for control of disease or full body health. Does the term "spontaneous remission" give you any ideas? Did you know that this phrase represents the Human Being who consciously or unconsciously gave intent, usually in a life or death situation, to claim the cellular structure of a past life that did not have the disease? Think of this! These are the new gifts that you carry around with you, given to you by the Pleiadians.

Chapter Nine

ROCHEV BAARAVOT is the Hebrew *name of God* associated with this Layer Eight. My meaning for this is "riders of the light." I have never revealed why this is my meaning, for it is a solid wink to the Pleiadians who allowed for you to have such DNA. They shared their lineage and wisdom with you, giving a gift that took thousands of Earth years to settle and become mature. They are, indeed, the ones who have "ridden the light of the stars" to bring you into this place of divinity. There is also something hiding here, for you ride the light. The very Hebrew word Merkabah is "to ride" and we will give you more about that in this book later.

There is also a Lemurian name, just as the other Lemurian layer. That name is **AKEE YAWEE FRACTUS,** and it means "wisdom and responsibility." This is very Lemurian, for here is the first large society in existence who knew they were different, somehow spiritual, quantumly endowed [interdimensionally aware], and responsible for finding spiritual purpose. It is these who gave wisdom to the others, and who planted the seed of knowledge that would eventually spread all over the earth.

Today, most Humans on Earth eventually search for the creator as soon as they are old enough to do so. This is an attribute that is caused from "missing the other side." You know intuitively that you are part of God, that your soul essence isn't just biology. It causes even the atheist to cry out to God the instant before he leaves the earth on his deathbed—sometimes in solitude, sometimes not. For it is imprinted in his DNA, within the 12 energies, that God is inside and is related to him.

The number *eight* is filled with misunderstanding. The meanings that are given in simple numerology are responsibility, structure, practical things, manifestation. Many look at this list and it looks entirely like a list their parents gave them! [Kryon humor] Instead, it is a beautiful list of how the Akashic Record manifests within the Human duality. [From Lee: *There is that Layer Two again!*]

Akee Yawee Fractus sounds a bit like a formula, does it not? It actually is, for in Lemurian, it represents a saying that they had. I have never given this before and I waited for a more full explanation within the pages you are reading. The saying they had was posted above the Temple of Rejuvenation, a place where they were able to extend life with the magnetics of the geology around them. The saying was: *"The solution to your problems lay at the feet of practical things."* This saying is not that elegant, for remember, this civilization was not yet that wise itself. But the saying was an axiom of wisdom that encouraged those who saw it to take care of the practical things in life before esoteric things.

It is the forerunner to many of the ideas of today that indicate that if you "know yourself" first, then all will be well. This is not high-minded, esoteric thinking, but rather it is practical thinking that contains the idea to "look inward to solve outward issues" and was one of the core beliefs of the Lemurians. It also pushed them into action and invention quicker than other societies to follow, for they viewed the practical as something very much aligned with spiritual values.

In Lemurian, the word *Yawee* appears in both of the Lemurian layers. This is because it was a placeholder for the word "wisdom" in their language. Anytime it occurred, it meant that whatever followed would have wisdom connected to it. It was also an honoring of the great Lemurian scientist Yawee, who gave them the technical design for almost everything that healed them.

The Lemurian language was a combination of linear words with a syntax construct (as you are reading now), combined with "word symbols" such as Yawee. Therefore, you couldn't read it in a linear way, as you are doing now. It was more conceptual. Eventually, the word symbols became only marks, almost like a shorthand of sorts, and Yawee was one of those marks. There are languages you use today that continue to use the Yawee mark. Look for them, and smile at the revelation of where the mark came from.

The number *eight* working with the number *two* creates the number *10*. Again, we have new beginnings! [Ten becomes a *number one*.] This is another signal, is it not, that the numerological aspects of these layers tell a full story of their attributes? The numerological story is, therefore, one of *wisdom and responsibility*, which creates manifestation of *new beginnings*, all wrapped around the Akashic Record of the Human Being.

May it not be lost on you that manifestation is the primary element of the *4*. Whatever the other concepts are (responsibility, practicality, structure), together they are a creative energy and that is manifestation. They manifest wisdom and this eventually creates a peaceful Human Being. What this layer is really about is the key to the Human puzzle—how to solve the blockages inside you and move into a more peaceful life, where karma is no longer an issue and other Human interaction does not create anxiety. It's the energy within the esoteric Human soul that begs to be examined, and promises solutions even to disease and unbalance. The Akashic Record within you is one of the greatest tools you will ever deal with.

SUMMARY: Layer Eight - Lemurian Layer Two

This layer represents the most powerful spiritual tool the Human has, and it is the personal Akashic Record of the core soul. Every lifetime is there, and all setups are there. It is interdimensional and not actually present in the DNA, but rather it represents a dynamic instruction set as a direct link to a far larger and more profound set of energy information that is always available in the Cave of Creation. Interdimensional attributes do not have location or linearity, so this is not something that is easily understood. We say they are in a "quantum state," where the smallest parts are transferred to all of creation, all together as one.

It is the second Lemurian layer and was part of the divine information given through the interaction of the Pleiadians at the beginning of the seeding of a spiritual humanity.

LAYER NINE: The Healing Layer (see page 329)

This layer is *Lemurian Layer Three*, The Healing Layer and represents a multidimensional process within the Human body. Some of the layers are conceptual; others are seemingly portals, while this one represents a process or action that the Human body creates within DNA that is often seen as mysterious. It is a Lemurian layer, but not Pleiadian. This means that the Lemurians used it as a staple of Human healing.

In the classes that my partner teaches regarding this information, all the ears perk up at the revelation of *The Healing Layer*. Humans can't help themselves, and their initial reactions are always 3D! [Kryon smile] They take a look at the name and the meaning, and they instantly are on a quest to "activate Layer Nine!" But it isn't what you think.

The double-helix, protein-encoded parts of the DNA carry the instruction sets for the genes. We told you that. Within the genes are complex coded instructions on how they perform, where they are in the evolutionary scheme of humanity and the chemical 3D reactions that control immunology. Then the white blood cells race to the place where the cut has occurred in the skin; this *is not* the workings of the ninth layer. Even when the body's full defenses are alerted to a virus that is killing it, the ninth layer is still not the one involved. All that I've just described is chemistry and a fine system of immunology that is part of the body's 3D existence and survival scheme. But it's only half of it.

Chapter Nine

Like these other layers of DNA past the first one, this one is also multidimensional. It works best with layer one [the 3D, double-helix, protein-encoded chemistry], since it is fully involved in the healing scenario of a 3D existence, but it is not the chemistry of the genes. Instead, you might call it a "knowing, healing, innate body consciousness."

In ancient Chinese medicine, the energy meridians of the body are addressed. Are they addressing the immune system? Ask those who administer it. They will tell you they are, but indirectly through the innate energy awareness of the body. Dozens of systems on the planet, ancient and new, all involve working with "body awareness," and that is indeed working with Layer Nine.

Did you ever wonder about the core of the body's communication system called kinesiology? There is sometimes called "muscle testing," where the body gives signals of acceptance or rejection to health questions put to it. Sometimes a person will hold a substance in his hand and ask the body if it's acceptable to ingest or not.

This is often seen as hocus pocus by science, for their world is allopathic and works with a defined chemistry of set 3D rules of chemical interactions that they continue to discover. What they are missing is before them as the looming *elephant in the room*. Somehow the body "knows" what is happening, way beyond the chemistry of the blood or even the cellular structure. This is evidence of DNA Layer Nine, the layer that is within the body and is designed to create healing of the system, but not in a linear way.

Layer Nine responds to multidimensional energy! What is that, you might ask? Well, my dear Human reading this page, how about the most elegant one: Human consciousness? Throughout Human history you have seen the results of this, and you call it by many names: Prayer, meditation, worship, faith, and positive thinking.

The results? Miracles, spontaneous remission, and total, complete healing. Are you starting to get the picture here?

So you might say that there are really two immune system branches in the Human body. One is the 3D chemistry, which provides a first defense, but needs to work with Layer Nine and often can't. Now the revelation: Layer Nine is a multidimensional layer in every piece of DNA, but it lays there and does nothing unless the Human Being or another multidimensional energy somehow *talks* to it. This second immune system portion is profound [Layer Nine], but it is not bound by chemical rules. Instead, it can alter the very magnetic portions of the cellular frame, creating a kind of healing that is not understood, seems like a miracle, but which is multidimensional and more powerful than any chemical reaction the body has.

"Kryon, I don't understand." Some are asking right now, *"You mean that we have something inside us that can heal us completely, and we don't know about it?"* Yes. That's exactly what I'm talking about. This second portion of the healing system of the Human Being is one that must be discovered through belief, or activated through certain kinds of multidimensional processes. It can be consciousness, energy systems, or (get ready) the devices being discovered on this planet that carry a multidimensional information instruction-set that quantum cellular structure will react to.

In Lemuria, they had something you do not. They were "fresh" from the seeding of the Pleiadian culture. Multidimensionality was a way of life, and they had no problem understanding something in a quantum state of seeming randomness. Were they advanced scientifically? No. They had no computers or telescopes. Yet they knew about DNA! They knew about the solar system and even of the galaxy. Did you ever wonder how so many of the ancients knew of galactic motion, yet when modern men with modern tools

Chapter Nine 177

started travelling, they were afraid they would fall off the edge of the earth? There was not even the acknowledgement that the earth was round! Seemingly, humanity had lost everything they knew about the stars and were starting over—and you might say that's exactly what happened.

As modern men became more scientific, they became more 3D. The ideas of quantum thought became separated into religion and spirituality. Some of it split off to become occult and even witchcraft. But the truth is that it was not seen as being solid science, for it wasn't predictable, couldn't be seen, and didn't make 3D linear sense. The ironic twist is upon you, for now it is science that is discovering the quantum state, where particles behave in a random fashion and can be in two places at once. It is science who has discovered that the Universe seems to be "biased for life," against all odds. It is science that will eventually discover the elegance of multidimensional energy, ironically with the help of those who are esoterically based, and who have spent years working with it.

So the Lemurians had their Temples of Rejuvenation. The "machines" in those places were not really machines as you know it, but rather "places of enhanced conscious thought." Isn't it interesting they called them temples? For later in history, you define a temple as a place of worship, but the Lemurians were not worshiping; they were dealing with the "temple inside." These were places of incredible, enhanced energy and information, all done with Human consciousness and magnetics in a process designed by a famous Lemurian scientist who outlived most of them. I leave it there, for we have gone over this before in publications given to you almost 20 years ago.

Therefore, the Lemurians were actively working with Layer Nine, which then "spoke" to Layer One (the 3D DNA double helix). The result was very long lifetimes, no disease and tremendous cultural health. Now you know: The multidimensional energies of your body are designed to "talk" to the 3D parts, thus creating a bond of consciousness allied with the genes, and the perfect marriage—a full-body experience. In order to fully understand this, you must believe that the body has a "knowing" that is far beyond what you see in chemistry. Then you have to honor and develop it.

You have lost this full-body experience, and most of humanity only works with the linear chemistry. Chemistry alone, without direction from the Human "boss," the consciousness of the Human, is like a boat without a rudder. It may or may not get to the destination (healing and good health) and may often stumble, go in circles, and most often fail completely, never making it to port.

Doesn't it seem odd, dear Human Being, that you have a chemical immune system that can't even tell you if you have cancer or a life-threatening virus? Instead, you have to go to a doctor for tests! What kind of immune system is that? I'll tell you, it's only half of one. The "knowing" part just sits there, not active and not working.

We told you more than 20 years ago that there would come a day when Human consciousness would play a part in creating a seeming "split" in who survived and who did not. It's beginning to happen. Over time, there will be a correlation seen of those who are starting to use the system of "talking to their cells" and living long, healthy lives. Against all odds, and counter-intuitive to 3D science, there will be those who seem to "think" their way to health. The facts will show it, and it may even lead to fear. For when one group lives far longer than the rest, they are often seen to be evil. Such is the old energy of Human logic, where drama and fear are the first things to be seen and weighed, even when standing in the presence of miracles.

Layer Nine reacts to multidimensional energy, and it's time for your culture to begin to also receive some multidimensional "health inventions." Why? Because it would be difficult and challenging for only an elite group of esoteric thinkers to be the survivors of humanity. It would, indeed, create a split, drama and, perhaps, even war. Like all things before, there must be a balance. So within those thinkers will come interdimensional inventions. They will help build two more rules of physics, and eventually allow for all six rules to be seen and appreciated. Science will someday breathe a sigh of relief, for within the six physics laws will be all the answers that have eluded them for centuries.

Humans from all over the globe will be able to take advantage of inventions that "talk to Layer Nine" within the scope of the "intent of the quantum invention." This is difficult to explain, but like Human consciousness, a multidimensional invention has the consciousness of the inventor and the patient imbued within it. Such is the nature of a multidimensional state where there is no time. Therefore, you might say that the intent of those who create and use the invention is then "seen" by the DNA. Although not near as effective as what might come from the consciousness of a Human with his own body, the inventions will work within the scope of certain situations. Most of these inventions will be to address disease that is not curable by 3D methods. It will not be the kind of subtle instructions that a Human would give to Layer Nine to create a long, healthy life. The inventions will address basic survival.

Let me take you somewhere for a moment. I wish to take you back in time to a real experiment that occurred on Earth, where the details must remain hidden for now. Eventually it will be seen as valid science and in order for that to happen, the protocol of your medical institutions must be honored and the system of development must be followed. All this will create something that cannot then be denied, for when mainstream science gets involved with multidimensional inventions, there will be eye rolling at first. Then the final "a-ha" will occur, which is the seed of discovery for

Human nature, for they won't be able to deny the numbers that will follow scientific protocol to the letter, and create validation over and over. It will be an interesting conundrum for them: provable weirdness!

Let me take you to the place where a dying man is laying next to a machine. The machine is taking the blood out of him, cleaning it, and putting it back in. His virus is killing him and the blood cleaning is only a stop-gap measure, for the viral load in his blood is too great to be eliminated by any method known to man. It can only be slowed, and not by much. He knows it. He is being attended by his doctor, who is administering the process.

He and his doctor have given permission to try something very unusual. Two men enter the room and set up a device that will shine a very thin light through the tube that is returning his blood into his system from the machine. The new device is large, and the power needed for it is great. The size of the device will change with time, but for now it is cumbersome and saps the electrical circuit of the building. All is set and ready. The blood cleaning begins and the device is started. It is a multidimensional device, the first one of its kind on the planet.

The blood cleaning is finished, and the patient is removed from the room. Some blood is taken for tests, as is typical, and he leaves. It's not very dramatic, is it? The little light that shined though his "returning blood" tube was a fraction of the size of the tube. Whatever blood was exposed to it was less than 50% of what was travelling through the tube. This means that there was no way that whatever the light did exposed all of his blood. Yet, the men were jubilant at the opportunity, for due to their former lab experiments, they already knew what might happen.

When the dying man's blood was tested, the results were not to be believed, for there was no sign of the virus—none. It had

Chapter Nine

gone from overload to almost nothing, reduced to such a small amount that conventional tests couldn't detect it. It was just as the men expected, and was outside the logic box of anything that was linear or 3D. That was the first time a live Human was treated for this killing virus in this way, and the first time that a quantum effect had been seen in real chemical tests on a dying Human with a system-wide virus.

If all the blood had not been exposed to the "healing light," then how did it reduce the virus from his blood so dramatically? This, dear ones, is the key to Layer Nine. This is the part where you should perk up and listen! For the body "knows" what the intent is, and the intent starts a process that is greater than the sum of the 3D parts used to create it! This is self-healing at its finest and it is not linear!

For the DNA "saw" the instructions... the information within the multidimensional "signal" within just a fraction of the blood exposed to the light. It started a process that is totally internal to cellular structure. It began self-correction. It enhanced the chemistry of the body, and then it actually altered the relationship of the magnetic structures of certain minute processes within the body in respect to the virus, and stopped the virus from attaching itself. The body "knows," and what it can do by itself is far, far beyond the 3D chemistry that you study, which you believe is "all that is." There is far, far more here than meets the eye, and this was the first experiment of its kind to prove it.

The man died, eventually, and was welcomed back across the veil. He was celebrated as one who came into the planet to allow such an experiment, and to play his part in history—a history you will see some day and one which will relate to this story.

His death was caused by those who followed 3D protocol, even after the virus had been beaten up severely, for it gradually returned and killed him. No follow-up visits were provided and the conventional curative chemistry only made it worse, for it didn't honor any of the Layer Nine attributes and this actually damaged the remaining immune protection.

Layer Nine needs to be addressed daily in an elegant way to create health in a Human Being. You see, it's not about reactions and chemistry. It's about the "knowing" of the body in a partnership with the sacred parts of the Human Being. These are the parts that are not seen or recognized by mainstream science, yet they see the results of it daily with those who *will themselves* to stay alive or clear disease from their body (spontaneous remission). There is nothing stronger than Human intent.

The invention could have worked, but it would have been needed a number of times, continuing to eliminate the virus until it could no longer survive in the Human. Even after that, the patient would have been asked to continue intending, another eye-rolling principal within this entire discussion, where 3D mysteriously meets another system that has no empirical rules of application, dosage, reaction or understanding in physics. Science is often this way, and it takes a long time to release what has been taught for decades and let something new begin to work and show itself as viable. This is exactly why we are not revealing the details of this first experiment—the when, where and who—for eventually all this will be known, and the physics of it will be recognized as valid.

The multidimensional inventions of the future can be extremely effective in clearing initial disease, abnormal growths (cancers) and viruses from the body. But the inventions can only carry "original intent," and as good as they will be, the Human will have to do follow-up with the primary quantum source—his own will and intention.

Chapter Nine

It may be confusing to the reader to understand the relationship of Human intention, pure energy work, prayer and meditation, or the application of a physical invention on Layer Nine. So, we will try to give you some insight. Remember the automatic transmission? The parts interrelate to each other. As one does one thing, it actually changes the roles of the other parts around it. No wheel or valve or gear within it can only be "one purpose," for it changes as it is being used in a complex scenario of ratios that respond to what the other is doing.

SHECHINIAH-ESH is the Hebrew *name of God* associated with this Layer Nine. My meaning for this is *The Flame of Expansion*. The Human Being who learns to use this healing layer and talks to his cells daily is kindling the flame of an expanding consciousness. It is a Human attribute that you consider a flame as something significant, even assigning it to the tombs of those who you wish to celebrate and remember forever. So it is with this layer, for it is eternal and unending and begs to be discovered.

We honor the great Human Saint Germain. His work continues to this day, for he gave information on an energy called "The Violet Flame," which is a direct use of DNA Layer Nine! This is the part of the Human body that St. Germain works with, for it is, indeed, multidimensional and part of an incredible healing process.

The numerology of this layer is interesting. It is a nine, so it represents "completion." The kind of completion it represents is the completion of the Human Being's ability to engage the second part of the healing scenario... a multidimensional process that compliments the 3D process in Layer One. When this is done, the Human is complete and is operating as designed. The rudder is in place and the biology is then being steered by the attributes that were originally put in place within the DNA to create long life, health and a system that "knows" how to heal itself.

Again, when you then combine the nine with the one, you get a 10. The 10 is then a "one" in numerology, and this represents new beginnings. Isn't it interesting, Human Being, that this healing layer centers on the beginning and the end? Start and completion are the two polarities of balance, the alpha and omega of the Human body. This is the balance we have spoken about.

SUMMARY: Layer Nine is the "missing part" of Human healing. It is the Lemurian Layer Three.

Way beyond the chemistry of the immune system, it is an energy within the body that "knows" what is wrong and that can alter cellular structure at the quantum level to provide tremendous healing power and stability. No chemical reaction in the body can even come close to this, since it is multidimensional and, therefore, can work far beyond the linear ideas Human have of what is possible. It is the miracle healing energy of the masters and relates to St. Germain and the Violet Flame.

It lays dormant, waiting for interdimensional "signals" from sources that vary from Human intent (consciousness), pure energy work from outside sources, and some new mechanical inventions that humanity is currently working on. It completes the scenario of the Human's ability to heal himself and continues to live without disease and with very long lifetimes.

It is the third Lemurian layer, but belongs exclusively to Humans, and not Pleiadians, for it was Humans who learned how to use it for their own bodies, giving themselves the ability to rejuvenate, cure themselves, and move forward in health. The activation of Layer Nine was the purpose of the Temples of Rejuvenation in Lemurian times.

Chapter Ten
DNA GROUP FOUR
Layers Ten, Eleven, and Twelve
Kryon
Book 12

Chapter Ten
DNA GROUP FOUR
Layers Ten, Eleven, and Twelve
by Kryon

DNA GROUP FOUR: The Divine God Layers

The God layers are Layers 10, 11 and 12. The interdimensional attributes are now at the maximum, and these three layers are not in 3D at all. In fact, many will wonder what these actually "do." For now, dear one, you must get used to some of these layers just "being" instead of having a specific function. These are not processes or memories as the last layers have been, but instead think of them as the pot that the meal is cooked in. Part of the process of eating and sustaining yourself is the kitchenware. These layers represent the "divine couch that DNA sits within," and is needed for others to exist.

LAYER TEN: Divine Belief - God Layer One (see page 330)

You might think that divine belief, that is the "call to understand your own divinity," would be a process or an energy that you draw upon to help you get through life. There is some truth to this, but the other layers we have discussed are filled with this call. The Higher-Self (Layer Six) is the conduit for discovery. So, what's this layer really about?

This is a fully quantum layer, like the next two. That means to us that it has the attributes of a true quantum state. What I'm about to say is not fully understandable by your single-digit-dimensional Human mind, but neither is a quantum state!

Chapter Ten

Let's say that you are an artist, and you are about to paint on a canvas. But this particular painting is going to have a background flavor to it, so you begin by painting the entire canvas light blue. Everything you do from then on paints objects within the "couch" of the light blue background—a constant flavor of color that helps you capture the energy of everything else you are about to paint.

Then you paint objects: the clouds, the fields, the flowers and trees. You end up with a 2D painting within a frame that is complete. If you ask the artist, "Where is the blue?", he will say, *"Everywhere. I started with it."* But, you might say, *"Yes, but you covered it up with other colors when you painted the objects."* The artist will then tell you, *"Yes, but it's still there... under the others, so there is still blue there even though you can't see it."*

The metaphor here is that your DNA must have the God layers to posture or begin the painting of the rest of your divine blueprint of life. So consider these layers as the background color. Although it is at the end of the DNA layer numbers (10, 11, 12), we have told you that there is no linearity within this teaching.

So at this moment, we invite you to create a circle with the numbers and create a golden DNA ring in your mind. This ring is an unbroken circle of one through 12, and when you glance at it, there is no obvious beginning or end. So you end up with a ring segmented into 12 parts, but no start or finish—therefore, no linearity. Now you are going to melt the ring and recast it, thereby blurring all the segments together. When that is done, there is no way to identify any part or segment or layer. This, then, is the way real DNA works outside this *classroom*. So, where is this specific layer? The "God Layers" are like this; they are everywhere in the DNA spectrum and simply cannot be segmented. But we do anyway, so you will know what colors the canvas was painted when the artist started [Kryon smile]. *"Yes, but if you include three background God layers, then you painted the canvas with three colors. Which one is the background?*" Ah, they all are the background, for they were painted all at once. Don't figure it out, Human; just celebrate it.

A true quantum state includes particles that are said to be in an entangled state. This is the only way to try to explain something that is "everywhere and nowhere at the same time." It is the basis for many of the quantum actions that appear to be random and not linear. When you have attributes that are everywhere at the same time, you can no longer separate them, count them, or interface them as you have in the past. All your linear rules of computation and/or geometry do not support something being "everywhere."

The belief layer is like that. It touches all the layers when it is needed, and it is, indeed, very needed! Did you ever wonder what it was that led you to understand the esoteric attributes within your life? As you became more and more familiar with the spiritual concepts of the Higher-Self and started to co-create situations in your life using synchronicity, the more you moved away from a truly linear life, and the stranger you appeared to others around you!

Many who are very spiritual and who love God as much as you do moved away from you since everything you were doing seemed out of place, strange, weird, and perhaps even unbalanced. After all, if you decided to start doing mathematics by painting the numbers in the sky with an invisible brush, then changing them randomly during computations, the math professor would probably ask you to leave the class! Many of you have been asked to leave the class of life because you started to become more quantum in your thinking—the very thinking that we have asked you to learn to understand.

God is multidimensional and Humans are not. This has been discussed many times throughout the Kryon work, and my partner sees it every day. He even sees it within the very new age halls he presents in, for there are those who are stuck in what I will call old

3D metaphysics. There they are, channellers and healers, sitting in a meeting trying to comprehend what is being taught, this very teaching. But instead their minds are objecting to all of it, saying to themselves, *"What's all this science? What happened to the good old days of just floating along, channelling when we wanted to, long deep meditations, and lots of good, strong energy work? What happened to meetings where we would spend three or four hours in meditation, sending energy to the planet and each other?"* Well, I have the answer, and many of you won't like it.

Your spirituality is evolving. Why would you spend hours looking into each other's eyes or chanting mantras when every bit of energy you create can now be done in the blink of an eye? Your very time frame has changed! Could it be you are stuck in a system you felt was static and forever—and very comfortable? Do you really believe that peace on Earth at this point in time will be created by those who sit in small groups and spend hours in an alpha state? Think: This is what the ancients did, and you are wishing to still be in that ancient system. Your profound spiritual DNA is evolving, allowing you to accomplish far more for the earth by becoming quantum in your life. You now possess the power of hundreds of chanting new agers who just want to sit and "be." You alone can affect the light-dark balance of the planet as your Merkabah moves from place to place with a very high vibration.

So, have I offended you? This was not the intent. Instead, it is to awaken you to a new concept—that Humans are becoming enabled to a place greatly beyond what the old views ever told you were possible. This is the study of mastery! This is you beginning to take on the energy of the ascended masters of the planet! You can't do that exclusively by sitting in a chair and meditating any more. There is a place for everything, and today's group meditation should concentrate on raising the vibration of the *individuals* in the room. Use it to help each other find the creator inside and become masters.

There is now a balance of meditation and manifestation and what this book is teaching is that the new metaphysics is evolving from the old. The very name *metaphysics* means "beyond physics." When today's physics has discovered multi-dimensions, you are no longer *beyond* anything as you sit there in yesterday's energy, longing to spend hours bathing yourself in bliss. Think of it this way. Take a bath, then go out and heal the planet! Don't just sit in the bathtub enjoying yourselves. That was yesterday's spirituality.

"Kryon, are you saying that group meditation doesn't work anymore?" No I'm not. It works just fine, but so does the horse and buggy. It will eventually get you there. But dear ones, don't you feel it? You don't have time! You don't have time!

VAYIKRA is the Hebrew *name of God* associated with this Layer 10. My meaning for this is *The Call to Divinity / Recognition of God in you.* This layer could also be called "The Divine Source of Existence." It is the first of the Divine God layers and represents the call to understanding of your own divinity. It's a "one" in numerology, which is "new beginnings," since this is the very first energy that a Human creates about the possibility of the belief in God.

The illustration within this book looks like the number eight. It is not. It is the infinity symbol, for this is the first time a Human must start a thinking process that has no beginning and no end. Look at the illustration, for it's also a rainbow. Did you know an actual rainbow is a circle? There is no beginning or end of a rainbow. It is not an accident that the artist saw this shape and color when it came to illustrating this layer. It is the profound belief layer, unending in its potential and, like the rainbow, cannot be contained, separated, or itemized. It is truly a quantum effect. A color is about to be painted on someone's intent.

So let's discuss something about this energy, this DNA layer. You all know people who are simply not interested in anything

Chapter Ten

spiritual. Oh, they may go to church and work through the motions of a protocol that pleases their family and perhaps even their peers at work. But when you speak to them about what you are doing, they either walk away, or if they care about you, they will say, *"It's simply not for me."*

Here is a Human Being who has not painted the background at all. There is no "couch" to build a foundation upon and absolutely nothing that will give him or her a reason to pick up a book like this. In fact, it's not real to them and slightly bizarre. Some would use the term "lunatic fringe."

Here, then, is free choice at its best. The "call to divinity" is not being heard, nor is it requested, nor is it anything that the Human really even wants. So understand this: Here is a layer that is a divine God layer, but which the Human picks up first. It's the background that gets painted so that other things will take seed and mean something.

Joe attends a funeral of his best friend, Mark. They are miners and the friend didn't even reach 30 years of age when he was killed in a mine explosion. Joe sits alone before the coffin pondering life itself. When the minister or priest comes in, Joe is open. His heart is warm and soft, and he wants solace since he is hurting. Mark is gone and he can't believe it. Perhaps the priest tells him about a very loving God and that Mark is there now… can't Joe feel it? Perhaps the minister tells Joe that life is a circle, and that Mark is still with him, but in a different way. He thinks about it and, indeed, Joe feels it for the first time in his life! He has allowed a spark to ignite a flame called *Vayikra* that is willing to work overtime to let this precious soul find a new truth. This truth will not leave him between Sundays. Instead, it may start painting a background that someday leads this man to ask, *"Sir, please tell me more."*

Joe, the miner, leaves the funeral and thinks about what the priest said. He does, indeed, feel his friend around him. He goes to the family and cries with the widow and consoles the children. He openly weeps with the thought of—yes—that God may actually care and that Mark's energy is still there. It gives him the desire to move on! He smiles at the thought. BANG! Layer 10 has just activated itself in his life through the compassion of a minor's heart. It is not doctrine. It is not angelic. It isn't any of those things. Instead, it is painting a background so that whatever the minor decides is spiritually correct for him, there will be a belief layer—a setup to help him create his faith, no matter where it takes him. He created it, and it began to assist him in his quest.

Blessed is the man who finds the creator inside, anywhere, in any circumstances, to any degree—for this is what changes the vibration of the planet.

This energy is, therefore, one that responds to spiritual intent, questions and puzzles. It is there to assist those who ask, *"Is there really a God? I want to know."* Those who ask it in an academic way will get an academic energy, not the belief layer. This one responds to pure intent and compassion.

SUMMARY: Layer 10 is the first of the divine God layers, and actually the first multidimensional layer that begins a Human's search for God. It is hooked to free choice and does absolutely nothing to convince a Human Being of anything spiritual unless it is asked to. Then it simply allows for learning, and still does nothing to teach or convince. Instead, it helps posture grace and understanding, allowing the Human to discover on his own.

Like others, it lays dormant, waiting for interdimensional "signals" from sources that vary through synchronicity. It is the first Divine God layer, and does not awaken fully until the Human asks for it, recognizes it, or is in a situation where he needs it.

Dear Human Being, this is the love of God at its best, where there is an energy in your DNA to help you believe! It works with free choice, but once you decide to look for the creator inside, it begins that background paint job—something that will help everything you look at make sense as you move forward in spirituality.

Today, Joe carries Mark's ring with him into the mine shafts. He wears it on a necklace around his neck. He smiles at the thought that his friend is with him in whatever way God intends. He is the godfather of Mark's son and proud of his faith in God—something that he is unashamed of in front of the other men who are younger. He kneels in church and feels the love of God pour through him. He is a spiritual man in his own way, convinced that God is real and alive inside him—all new due to what happened to Mark.

Here is where we might close this layer, but we didn't tell you anything at all about Mark. For within Joe's DNA, Mark is there, connected by the "background that is everywhere." When Mark took his last breath, he gave intent for Joe to discover the truth. So you might say that Mark's death, tragic as it was, created a life of its own in Joe. And you would be right. For this is how it works, dear ones, that out of the tragedy of death and heartbreak often come the epiphanies for many on the planet, so that the relatives and loved ones may indeed activate this God layer. Therefore, compassion wins again, and the process of Human free choice to find a loving God often trumps any and all sorrow or death, for the one lives on in the hearts and minds of the many. Meanwhile, Gaia herself raises her vibration because of the many *Marks*—everywhere.

LAYER ELEVEN: Wise Divine Feminine—God Layer Two (see page 331)

The women will smile and the men will roll their eyes, for here comes another lesson in Goddess energy, so they believe. Well it isn't. It's a lesson in balance, and one that the Human race has not had since the Pleiadians arrived and gave it to you. This is the *Pure Compassion and Mother* layer.

Again, it's a multidimensional layer that is everywhere at the same time and is part of the God layers that do nothing unless you call on them. It's in an entangled state with the other layers, which is the fact that it's truly quantum and cannot be seen as an individual energy.

Here is the truth: Human nature has been unbalanced for eons, and the shift before you is starting to make a difference in the male/female balance of the Human Being. Your consciousness is not controlled by the brain or the genes. It's controlled by the sub consciousness of an Akashic experience. How did men behave and how did woman behave in history? So, therefore, what is your role in life? It begins with the blueprint of information contained in your DNA as energetic history. Then it either evolves or de-evolves as your free choice carries you into uncharted territory.

The gender issue of humanity hasn't been balanced for a very long time, and since the males are larger, they win! It's about the most elementary and unenlightened feature of humanity, which never moved from the animals they saw around them—except the animals shared more than the Humans did! You might say that humanity continues to be at the lowest point of male/female balance that it ever was.

Layer 11 exists to help balance the compassionate Human Being so that the wise, divine feminine can show through both men and women. Your history is filled with it, yet you still don't honor it.

Chapter Ten

I will let my partner paint this story, since I'm not that good at cultural humor and he thinks he is. So I turn this section over to Lee before I return.

A Funny World:

Imagine for a moment, a society where women are totally in charge. This is imaginary. I'm not giving you a historic lesson—just a "what if." Since they give birth, there are very few wars. Their sense of motherhood keeps them wiser in the ways of settling conflict, so they never send their sons to die. Most of the taxes are collected to make shoes—for women, of course. Men just wear one pair and are happy for years until they wear out. Then the new ones should be exactly like the old ones. There are entire factories where men make shoes for men—all the same kind and color.

Women run business in an interesting way, with everyone talking at the same time in board meetings, lots of tears and passions running high. It creates noisy meetings but with solutions evolving out of honesty rather than the age-old system of secrets and never letting your partner know what you are thinking. Much more gets done. More shoes are made.

Men are larger, so they do the hard work. It makes sense. They do most of the hard labor. After all, why would they be so large if that was not their purpose? Clothing the men is not difficult. They will wear the same shirt for months before they wish to change, usually due to the unraveling of the cuffs. Then, of course, you just give them their favorite shirt again, much like the shoes.

Sex would totally be for a woman's pleasure, for they would call upon a man at any time and he is always ready—hence the chemical difference between them. God made men that way so that when a woman needed to prepare for birth, he wouldn't have to be prepped—always chemically ready. The woman's chemistry, being more complex, needed a more elegant timing. So the man's job was obvious—always ready so there was not a mismatch of attitude. The woman's pleasure would be multiple and she might need several men to complete the process! Such is the way men are, only good for one time around. Men are like that, they exhaust early. A good woman might exhaust several men just to get to the place where she is seeded properly. This creates magnificent children, with the best seeds reaching their mark, especially for making really good females. Females would proclaim that sex was actually made just for them!

Early on, the women would teach men the ways of cooking and cleaning, which women know all about, being the superior gender. Then the men would do their best to comply and follow their instructions. Interim courses would be needed along the way since the men don't like to open the instruction manuals very often and forget important nuances of seasoning and cleaning chemistry. But with repeated training, they do well. You have to watch them, however. They are distracted with primary animal contest instincts, dressing up like gladiators and pummeling one another.

Occasionally, a man will think so much like a woman that he excels to a political position and actually is considered for election. But women, who control most of the voting, try to keep that from happening too much. After all, a man should know his status and never get to the place where he earns what a woman does. God forbid he should ever reach high office. Think of what that might mean for the shoe consortium.

I think I let my partner go a little too long. Now he will get letters, but knowing him as I do, he has someone else read them first and burn the ones that are negative for him. So go ahead and send him whatever you wish, for it helps to heat his house and he never reads them.

Why did I allow him to use his humor like this in a way that might be upsetting or offensive to one gender over another? The reason is that his rather fun, but obviously demeaning picture of men, is what many women of the world are faced with today—in fact, much worse. Do you really think that the gender difference was supposed to be this way? Do you think that the Pleiadians were this way?

Let me give you a huge hint at how gender balance was designed to work. Those who seeded you were called the "seven sisters." Most of the mythology passed down for more than 100,000 years written in stone in many areas had women landing on Earth and small men chasing them! Actually, the men were not small at all, but the women were large. It fell upon the women to travel into space, for they were better suited for it. The Pleiadian men came later, and they had their specialties also—a very honored and shared culture, featuring a divine, wise, feminine energy. Both were scientists, both honored each other, and there were no barriers of gender in their culture—no stereotypes and never were there slaves as you have today.

All this is to tell you that within your DNA is the blueprint of how it should work. Again, it is an energy that just sits there until you want it. But in the new shift, it was activated slightly and brought to a more poignant energy with the Venus transit of 2004. Unbelievably, it happens again in June 2012! Usually these transits are hundreds of years apart, but these are very close together. The reason? Your planet needs this compassionate, mothering energy, and that's exactly what it carries to the earth. It carries love and beauty, compassion and softness, and it's only eight years apart. Again, the eight represents a manifesting energy and this is needed for the planet.

So, here is Hank. He's a typical guy on his way to work at the mill where his father worked and his father's father worked. Tell him about the Venus transit and he may walk the other way! Hank isn't interested, and neither are other Hanks by the millions on planet Earth. So what's the point?

This is a Gaia energy infusion. It's like tanking up on the fuel of the future so that when humanity is ready for it, the energy will already be here. Gaia and your DNA field are very much intertwined. But you knew that, didn't you? It's your DNA vibration that "speaks" to the Crystalline Grid of the earth, and that just has to be a Gaia agreement, doesn't it? The relationship to you and Gaia is so strong that one Lightworker can keep the earth from shaking in a certain area! Did you know that? So this entire layer of DNA is poised for the future of humanity and ready to activate as the wisdom of the masses begins to increase due to the higher vibration that is present. But it doesn't tank-up in Gaia. It does it in you!

Will you have what my partner described? No! Instead, you will have a culture who leads the way—not with women's rights, but with gender equality. It will be natural and feel correct. There will be no "glass ceiling" for earning power or gender-specific jobs. It's already happening now, but without the respect that will come from inside.

Yes, there is a layer of your DNA that sees the gender issue from both sides, since an old soul has been both genders! It calls upon the best of both and gives you the feeling of being both and not just one. This creates a feeling of ultimate respect for the gender that you are not this time around, and helps you to make decisions based on this. Imagine a man who has the "feeling" of childbirth or the woman who knows what it's like to line up on the battlefield. It's all *in there*, dear ones, and ready to be balanced in a way that is natural, normal, and pulls from the Akashic Record of all the lifetimes of old souls on the planet.

CHOCHMA MICHA HALELU is the Hebrew *name of God* associated with this Layer 11. The Kryon meaning for it is *Widsom of The Divine Feminine*. Let not the 11 number be wasted on you, for not only is it a master number that means "illumination," but it carries with it the number of Kryon. It is as I told you more than 21 years ago in the little white book—that my numbers are nine and 11, completion and illumination—the completion of the old and the beginning of an illumination of the shift.

Look at the illustration, for in the middle there is a flower. It's the only illustration of this kind. It's soft, inviting, and welcomes you to share it. The only exception is the last one, and it is a flower also—the flower of life.

Let me tell you a story, my dear Human. It's about a man who may not have been your spiritual leader, but he controlled almost a billion people of the earth, spiritually. Karol Jozef Wójtyla was born to the earth on May 18, 1920. His mother died when he was nine, something that affected him for life. He became Pope John Paul in 1978 and greatly influenced what happened on the planet from then to his death in 2005.

His spirituality was an oxymoron. He worshiped the "mother of God" as Mary, mother of Jesus, yet his system never let women be leaders in the church he was in charge of. As it was, this is what occupied him for much of his life. For when he was troubled and in pain, he fell to his knees and called upon Mary. It was Mary who got him through the years of pain and suffering that the church never told you about, and it was Mary and his mother who met him at his deathbed and walked him to the light and to the love of God.

As Christ's vicar in charge of Peter's church, you would think that these things might be different, but this man loved a woman named Mary. When then-president Ronald Reagan was on his way to Moscow, he consulted with Karol. The pope looked him in the eye, called on Mother Mary, and said these words, *"Trust the Russian president, for he wants what you want."* No cabinet member had told Mr. Reagan this. Reagan was ready to negotiate hard, as he had with all Russians. Instead, he was taken back and he pondered the advice of a man who wasn't even within his religious belief. Later, he took that advice, and today the results are obvious. The divine feminine in a Polish man named Karol did that.

Did Karol stop a world at the brink of war? Or did the advice from Mary do it? The answer is neither, for the Pope had his DNA Layer 11 activated at the highest level. He was a respecter of women and cried for their plight on Earth. Let history show that he even met with those in charge of the Catholic policy, begging them to at least let women become priests and perhaps eventually cardinals. It was turned down by the German who kept the old ways sacred. His name was Joseph Alois Ratzinger.

It is Pope Ratzinger who will be responsible for taking the Catholic church to the brink of destruction and even bankruptcy. He won't last long, however, and the new pope will be one of the most compassionate the church has ever seen, able to change the rules so that the love of Christ can again relate to the common man. He will save the church, as the potentials read at this moment. He may even involve women for the first time.

The difference in all the masters on the planet is DNA Layer 11. Can you see the feminine in the masters who walked the earth? Never mind their gender. Can't you see the balance? Can you see the compassion in the eyes of the men, the strength in the eyes of the women? This is a balance that is sacred, divine, and is helped by the God Layer 11 within the Human Being. It's there for that purpose, since gender can cause division, hatred, bias and worse. It's God's way of acknowledging that there are chemical issues at work that can be voided with wisdom, divine wisdom.

Where do you go on your deathbed, dear one? Since many of you don't remember, I'll tell you. You go to your mother, first! She is there for your last breath just like your first one. She may be long gone, but she circles the bed and gets ready to greet you first. The one who birthed you and was connected to you by the cord and her blood, with all the angels celebrating, is the one who welcomes you first. She is beautiful, seemingly in Human form so you will be peaceful and recognize her. She waits and waits and then is there for you. Her smile greets you and the reunion is one of the most beautiful events on Earth.

On all the battlefields of Earth, the mothers were there to meet their sons. Some of the men cried out for them and others just reached out to embrace them. With all the blood and dirt, the mothers lined up to be with their sons—on the beaches, the fields, in the mud, under the sea, and in the swamps. The divine feminine is one of the strongest forces on the planet. It is the commonality of death and birth and represents the compassion of God.

Intellectual, don't analyze this away into a metaphor. You might say, *"Well, she reincarnated long ago, so she obviously can't be at your deathbed."* Enough of your 3D arguments for a quantum God! Know that God is far bigger than your silly arguments in the box of your own restricted thinking. She is there, and you will know it when it happens to you.

SUMMARY: Layer 11 is the second of the divine God layers, and actually the most powerful force in the Universe, for it represents the *compassion of all humanity,* which is epitomized within the *mother* energy. It's an energy that isn't there unless the Human calls for it.

Even women don't have it unless it is called upon. It's the divine feminine energy and is for all genders. It represents "mother" in all of you, but not just mother, but a divine compassionate mother. It is the very reason such emphasis was placed on the mother of Jesus, when so many other religions didn't even include a woman at all.

It is the layer activated with compassionate events on the planet, such as the death of Princess Diana, or the tsunami, or your 9/11 experience. It softens the heart and opens both genders to common sense solutions and peaceful times. It is being stored up within you, but also controlled by Gaia. Together, it will be released when the shift needs it the most. This won't be a catastrophe, but the birth of a new energy.

LAYER TWELVE: Almighty God—God Layer Three (see page 332)

Does this layer surprise you? It is the third God layer and is the essence of the creator itself. What is God? It is simply the word you use for the creative energy of the Universe and that which is everywhere, quantum, with no beginning or end. God is also within you, but not residing in you. God has split apart a piece that lives in you, which is also in an entangled state (part of God—part of you), and that makes you a part of creation. The "God in you" is literal. There is a part of the creator consciousness in every Human who has ever lived on the planet.

Arguments break out all over many cultures when we speak of this. *"You mean the greatest villains in the history of mankind had God in them?"* Yes. But it means they totally ignored it and with free choice did what they did. *"You mean that when they died, there was no punishment for what they did?"* Correct. Punishment and reward are Human concepts that create revenge or celebration and somehow tie up a package that isn't complete unless a Human gets what they deserve. This is a 3D concept that you paste upon God, as you do so many Human concepts. In your mind, God rewards and punishes, too. It's in all your "modern" spiritual systems.

You believe that in heaven, there is somehow strife. There are angels fighting with angels for power. There are hierarchies of control. There is even a fallen angel who is also somehow allowed to be powerful, and who is after your soul. There is revenge and punishment and horror and hatred. There are demons fighting angels, and God is somehow in the middle of this as a "decider" of right and wrong.

Does this really sound like God to you? It isn't. It's the Human brain trying its best to justify something that just isn't there in a multidimensional state. So it places all the Human attributes on God in the sky and the mythology of the ages is filled with it. Humans can't think past their own consciousness, so there is a phrase that says, *"You can't know what you can't know."* Teaching a dog the attributes of the incandescent light bulb is not possible. The dog just can't go there. No matter what you try, the dog can't learn that. If you could ask the dog how it was doing, it would state, *"I can learn anything! Try me. I can fetch a stick and roll over. I can even learn to understand Human hand motions and emotional states. I'm a dog and as smart as it gets."* The intellectual Human has the same argument. It thinks it's unlimited in what it can think about, yet it fails completely when faced with things outside its own reality.

Tell me, dear one, can you think of something that has no beginning? No, for your truth is that everything has to have a start. In 3D it does, and that's your reality. Even science stretches and stretches to create a beginning of the Universe, when in actuality it always was; it just changed dimensional properties, as it does quite often. There is no beginning of time. There is no beginning of the Universe. There is no beginning of YOU! You always were and always will be. Yet the Human brain objects.

In a multidimensional state, there is no such thing as a "place." Can you think past that? No. Everything must have a place. In fact, your very geometry falls apart without a place to start and end. This is why even your circle is defined as an octagon with an infinite number of straight lines. You can't even think outside of straight lines! In some ways, your reality is as limited as a cartoon figure on a page, who has no idea about who the artist was that drew him. He can only think about what is on the page.

Time is in a circle, but you only face forward. Your 3D time features a forward bias, with nothing happening in the past, since it's static [already happened]. What if I told you that this is a fallacy! Time is reactive upon itself. That is, even what you call "the future" affects the past! The future is not known, but the strongest potentials of what might be on your timeline are, indeed, known. Therefore, those potentials affect what you do today. It's the future affecting the past. If time is in a circle, then the past affects what you do next. Those who study the fractals of time know exactly what I speak of and even have some basic formulas for examination of this phenomena, which is totally outside of your straight line box. Did you know that the Maya knew of this, too? They worked it out with their observatories and saw it with different eyes, with esoteric wisdom that showed the same waves of time affecting consciousness over and over. Too weird for you? Again, your weirdness is simply anything out of your intellectual box of reality.

The dog again: Did you notice the dog can love? Did you ever wonder if love was the basic truth in all the kingdom of all the animals and plants of Earth? Did you ever love a plant? Did it respond? Yes. Yet the plant can't even think. It only can react. Doesn't this tell you that perhaps there is something that is greater than even Human intelligence? What if the love of God is everywhere, in everything? Astronomers in the last 10 years have noticed it, too. Contrary to the way randomness works, they have seen that against all odds, the Universe is created for life. Everywhere they look, there is the potential for life to exist—everywhere. They have even given this wildly counter-intuitive natural attribute a name. They call it "intelligent design." There it is, an acknowledgement that somehow whatever created the galaxies themselves had intelligence and it is *biased for life*.

The creator is pure, unconditional love, biased for life, and a collective of trillions of what you call souls. There is no Human attributes on the other side of the veil save one—the love that you know is real, for it is also powerful on Earth. There are several things that pass through the veil untouched. They are love, music, art, creativity, and the desire for home. These are the only attributes that you have that are God-like. Everything else is of your own making, of your own invention, of your own 3D system.

This layer is responsible for the term "made in His image." It is this layer that is the image of the creator inside you. It does not mean that when you return to the other side of the veil, there will be Humans. This is very funny! No. Instead, it is the image of the creator that you will see, a very familiar family of multidimensional energy who is God. It has no time, no place, no mansions of gold and no pearly gates. It's better than that. It's home.

The Power of Humanity

Humans are powerful, very powerful. This part of them is what is seen by any life form who can detect a multidimensional state. One of the biggest ironies of creation is that you are in 3D, yet you carry within you such a magnificent quantumness that your quantum field is eight meters wide! Every Human has one.

Say, for instance, that a visitor from space came to tour Earth. They are familiar with quantum energy and multidimensional realities, and that's how they got here—the only way they can, by the way—for travel in a multidimensional state is almost instantaneous, since it creates an entangled state, as discussed before. This is an attribute that voids the whole idea of a "place" and where everything is everywhere. Someday science will discover how to program a 3D object to become entangled, then be "projected" to arrive in 3D, where it already is! Think of sound in a room that has no source. You can't identify where it's coming from, for it's emanating from every molecule in the room. It's everywhere! But somehow, through new science, you program it to suddenly be focused into one pin point on a table. Now you can only hear it if you bend down and listen carefully to that one spot. You have essentially temporarily captured it and put it in a "place." This is how your space travel will exist some day. Think about it. The vehicle can be as large as you wish, and you only have to be concerned with the environment of the planet you are going to, for the hazards of space won't even be involved, unless that's where you wish to go.

The space traveler is innocent. He has never been to Earth before. The first thing he does is to get out his multidimensional instrument and measure the energy of the dimensional shifts that may be present. This is critical to his survival, and he has to measure it to stay out of danger. I have just given you a hint of a far futuristic science attribute, but you can't use any of it yet, since you don't have the science to create the parts of the instrument.

He looks around and suddenly the instrument goes off the scale. Alarms go off, and he knows there is something very different on this water planet. The measurement is coming from the forest not too far away. He moves closer, but out of sight. There he sees his first Humans, having a picnic. They are beautiful! But the energy coming from them is absolutely scary, for it's the same kind of energy of a *singularity*—a black hole duality. It's different, but very unusual and dangerous to a multidimensional traveler, for his entire multidimensional engine is based on a carefully balanced, phasic displacement of dimensionality—controlled and very precise. It keeps him in the entangled state that allows travel at all. He must move off.

What he sees on his instrument is that each Human has an eight-meter field that is so powerful it contains the elements of creation itself. He has seen it before, when he travelled to places in the Universe where stars were being made. He has seen it in the center of each galaxy, where no life can go—and now here it is around each Human! The children have it, too. They all do! With great speed, he maneuvers his vehicle to stay out of the way of these creatures! He marks the spot on his star chart as "dangerous" life form and never returns. He ponders how such an energy can be controlled and contained within a corporeal body.

What a fanciful story, you might say. Yet it's real, very real. Let me ask you, dear Human, this question. If, as I have told you, the Universe is teaming with life and they can go wherever they want almost instantaneously, and you can often see them in the sky moving at speeds that are impossible in 3D, then I have a question for you: For more than 100 years of observing them, why haven't they landed on the White House lawn and announced themselves? Does this make any sense? For those who study ETs, you have all manner of conspiracy theories, don't you? They are waiting for something. They need a larger invasion force. They need a special galactic alignment, etc., etc. This is all 3D conjecture at its worst and is typical of the way Humans approach things that are outside of their reality. Fear is always first.

The truth is that this planet contains the ONLY creatures who have Almighty God inside them, and it's called DNA Layer 12. The creator is in you and you don't even know it. Your DNA field shouts it and it's powerful beyond belief. It can move mountains, keep the earth from shaking, stop volcanoes, and allow life for hundreds of years. But you don't believe it yet.

You are on the brink of discovery of what is inside you. This shift that started in 1987 and will continue for more than 36 years (measured from 2012) will set the stage of a time fractal that has been seen and predicted for eons. This is the reason Kryon is here. You had free choice to move into it or not, and you did. It has the potential for peace on Earth and this without any kind of mass conversion to a spiritual system, for the systems you have all over the planet will evolve and will start to include the wisdom of love and the integrity of the divine feminine. All the layers of DNA will be enhanced by themselves, just as they are beginning to be now.

A world without war does not mean that everyone believes the same thing. It only means that there is a wise, evolved consensus that war is an old energy barbaric system and that Humans just don't do that anymore. Instead, they use high-minded logic, outside of greed or manipulation, and the idea that the earth is entangled and countries are not outside of the laws of nature. They work together! You won't see this in your lifetime, but dear ones, you *will* be here. Some of you reading this, and perhaps even writing this, will help it to happen.

Back to the ETs, for everyone wants to know more about them. There are many kinds and many varieties. They have been arriving for years and leaving for years. Many never come back, for the reasons I have just given you. It's dimensionally dangerous for them here and they know it. There is one kind who may visit sooner or later, in love, and appropriately. Watch for the blue spirals, for they foretell the creative energies of Earth and who are now what the Pleiadians have become, for even they have evolved in 100,000 years into light beings. Do not fear them.

There is another kind, however, who are the most curious of the bunch. They keep returning and re-configuring their dimensional craft. None of their science can help, and they now realize what can void the power of the Human. They have discovered it and are using it even today. It's raw fear.

Fear disables almost every part of the quantum Human Being's power. It's a takeover and brings the Human to a basic survival modality. Many of you know this, for you have conquered it and know how it operates. Within moments, fear shuts down almost everything but the basic survival instincts. You can't remember your phone number, you can't cry for help, and you crawl up into a ball and shake.

These specific ETs are convinced that if they can capture Humans, mate with them (via their science chemistry), and completely understand the Human body, then they can achieve the greatest power available in the Universe. They can have what Humans have! So they try. They have learned what frightens Humans, and they count on it. Did you know that some of them even wear masks? This is funny, for it's like a horror show for maximum effect! Keep the Humans frightened and you can do anything with them—and they do.

So many of the stories of abduction are indeed so, and so are the stories of what happens in their ships, all done with fear. Fortunately for all of you, the power inside you is not available through any means to them, not even a biological mating. It's for Humans only, and no amount of science or fear will be able to capture that which is the essence of almighty God.

ETs see this power as mysterious. They have no idea what it actually is. This is also why we again give you the answer to how to keep them from coming back. If you will understand this scenario, then you can also understand that the attribute the ETs fear the most is that you would know what they are doing. So, for the few weak-hearted ones, we tell you to bolster up your courage and the next time they appear, instead of fear, shine your divine light on them and demand that they leave. They will have no choice, or be vaporized by your amazing power. They know it! They will leave, and you and your family will never be bothered again. It is unfortunate that the very Humans who need this information are the ones who in all probability will never see this book. For they are mostly in the country, filled with fear-based and conspiratorial information, and living lives that are simple and easy to manipulate.

EL SHADAI is the Hebrew *name of God* associated with this Layer 12. The Kryon meaning for it is *Almighty God*. The 12 in 3D numerology reduces to a three, and in our case represents the "third language." This is what we told you is the unspoken multidimensional language of Spirit that gives you messages through intuition, even while you think you are sleeping during a live channelling! The three is also DNA Layer Three, which we have called the Ascension and Activation Layer. These all work together, and are a catalyst for divine shift. As described earlier, DNA Layer Three works with DNA Layer Six, which is your Higher-Self. If I were teaching a course in multidimensional numerology, I would tell you that the quantum number associated with El Shadai is 33! For it is the energy of the three upon itself and "influence" counts in a quantum state.

The illustration associated with this layer is again like a flower. I called it the flower of life, but that's not official. It's just a phrase that indicates that this illustration is not abstract. It's something real and you can almost reach out and touch it. This golden flower seems to never end. Even outside of the frame, you can see it continues on and on, it's beauty larger than you think.

When Elijah ascended by choice, he was observed by Elisha, his understudy. As the story is told by Elisha, his master turned into an incredible ball of light. He appeared to "ride" in this ball, hence the Hebrew name that basically means "to ride." But the ball of light did not come down from the heavens to get him. Notice! The Human claimed his power and turned into the ball of light, which was his DNA field, eight meters wide. It was illuminated with all the glory of the multidimensional field that was Elijah. Elisha was astonished. His master turned into a ball of light and rode it into the heavens.

In addition, Elisha indicates that three white horses helped draw the ball of light that Elijah rode within. Since none of this was in 3D, Elisha did his best to describe the three, or the catalytic energy of the three, that is involved both in this ascension process. It was the Ascension Layer (DNA Layer Three), Almighty God (DNA Layer 12) and the Higher-Self (DNA Layer Six). This is the manifestation of the God within you. Elijah turned into the piece of God he was and joined the family on the other side of the veil, and the three divine energies that belonged to the other side of the veil took him home. (3+12+6 = 3)

The spiritual intellectuals, who are still trying to figure out what each layer "does," will scratch their heads with this one and eventually give up. *"Well, it just doesn't do anything!"* they will say. We can see how the others work, but not this one. It just sits there and exists. If you are looking for an "action" layer, this isn't it. If you are looking for divine life force, you just found it.

SUMMARY: Layer 12 is the third of the divine God layers and could be considered the energy of "God within you." One of the most quantum of all the layers, it truly is the creator energy within your own DNA. It has no function other than to exist and create the life force of God within the Human Being. It is always there, but like all the other multidimensional layers, it does not show itself until intent is used by the Human to discover just how powerful it is and how to use it.

Of all the layers, it is the one that is involved in all the other 11—such is the love of God that imbues itself into everything that is. Therefore, Layer 12 is a confluence of energy that represents the meld of God into all life force, especially the divine Human Being.

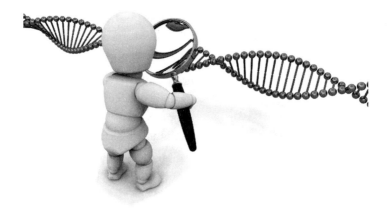

Chapter Eleven
THE DNA GROUPING SUMMARY
...and the big scary secret!
Kryon
Book 12

Chapter Eleven
DNA GROUPING SUMMARY
...and the big scary secret!
by Kryon

DNA GROUPING SUMMARY

In review, we have given to you 12 layers or energies of DNA. In the process, we also gave you four groups that the 12 belonged to. They are:

The Grounding Layers
Layer 1— The Biological Layer
Layer 2— Life Lesson Layer
Layer 3— Ascension & Activation Layer

Human Divine Layers
Layer 4— Your Angelic Name Layer 1
Layer 5— Your Angelic Name Layer 2
Layer 6— Prayer and Communication Layer

The Lemurian Layers
Layer 7— Lemurian Layer— Revealed Divinity
Layer 8— Lemurian Layer— Wisdom & Responsibility
Layer 9— Lemurian Layer— Healing Layer

The God Layers
Layer 10— Divine Belief Layer
Layer 11— Wisdom of the Divine Feminine Layer
Layer 12— Almighty God Layer

Your historic religious systems are far from having a sacred history. Any of you who study history will find that as soon as men discovered they could control large groups of people by becoming their temple leaders, an entire new process of control was born. Better than kings and queens, popes and shamans even had royalty at their feet.

History is history, and you don't have to embellish it to observe that spiritual systems created most of your wars and even today has divided more than it has brought together. Although today's spiritual systems have far more integrity than ever before, the historical side is filled with horror.

There was a day when the church rode across the kingdom on horseback and killed families who were not believers. Today they do it with airplanes. Popes were corrupt, and many cultures had a dynasty at the heads of faith that were more concerned with money than spiritual purpose. This is nothing more than Human nature in those days, and it's what is today being cleaned up and seen for what it is. However, even today there are the abominations of those who are supposed to be dedicated to God, and the search goes on to try and make the church a place that is without shame. This is honorable and part of the shift, as we have indicated.

Meanwhile, the secrets, even of the DNA, were known. They were kept as secrets by certain orders who today have names associated with conspiracy, or which modern fiction has degraded into mythology. But the secrets were maintained, and few knew them except for the indigenous of the tribes of Earth, who never wrote them down. *The ancestors of humanity have always been the key to spiritual truth.* Before large cathedrals, big money, huge projects or crowds of thousands, the shamans knew it and the wisdom keepers knew it. They didn't share them with anyone except their own, and now that itself is beginning to change as the shift is upon you.

One of the ironies of the planet is that the "New Age" is simply a study of the "Old Age" when the secrets of creation were still fresh and where the alignments of the heavens meant something personal. All of this is to say that not all your spiritual writings are accurate. Many are poems, songs, some laws, narratives, letters to friends and even political slogans, rehashed from very old history into the verses of our holy scriptures that don't apply anymore*. In your cultural religions, much of the book of Leviticus contains horrific rules that would have you kill your neighbor if he married your virgin daughter, or if you find a physic or wizard, kill him. These things are still seen as part of the package that is the "word of God" today, but usually ignored as something cultural from the past. Therefore, there is a loose acknowledgment that perhaps holy scripture was often written by men of the times, and not to be considered as sacred or infallible, or even literal.

Years ago my partner visited "the eye of the needle," which is a narrow gate in a wall in Jerusalem. A camel can't pass through this opening. Many feel the word camel meant "rope" in the original Aramaic, since they were spelled the same. However, I'll tell you what it meant, as it was translated up through four languages over time. The scripture that admonished that a rich man could not gain favor with God any more than you could put a camel through the eye of the needle didn't mean that a rich man couldn't reach God. It meant that a rich merchant would not be allowed into the court-yard of worship—to sell his wares. His loaded camel wouldn't fit! Therefore, the admonishment of the time was that commercialism and sacred teaching should be separated, and this is still true today. It's common sense, but look what the *men of scripture* did to it! It made people feel that they should give all they had to the church

---

Reference: *How the Bible Became the Bible* by Donald L. O'Dell
* www.kryon.com/Bible

in order to gain God's favor. This is manipulation, and is directly from men, not God.

The inspirations of Spirit are beautiful, and many were given in perfect harmony for the times. It's what men did with them that made the difference. Therefore, it tells you a lot about men, and not about God. Always look for the original, untouched manuscripts of the words of the prophets. Look at the cultural meanings present. These are the ones that glow with truth and the love of God.

All of this is to present to you the concept that not all of what you have seen as sacred, is really sacred at all. A great deal of it is simply history. Some of it is inspirational, and much is just the words of those in control at the time. And within that control there were admonishments to keep you away from the secrets, the very elements of sacred truth that would give you the information you received in this book, and which the ancients still have. For years, scripture wasn't even readable by the common man, and only in the presence of a spiritual leader. The common man had to go to others to interpret it for him. Can you see how this was ripe for control?

What if men had known what you know now? What if they had realized that their God was inside? Who could control them? What if they knew that they could heal their own bodies better than modern medicine? Would that upset anyone you know? What if they didn't need a church at all? Who would then have to pay the price? The answers are obvious, and it's time you knew a basic truth.

My partner goes all around the world. He carefully photographs it all so you can see the numbers of people who are drawn to a non-church event. He wants you to see that in every country he has visited, there is a shift going on. He places these photos on the Internet. Without mass advertising, billboards or TV ads, thousands are coming to these meetings. He presents one time, doesn't leave a team behind to evangelize, doesn't ask for additional money, doesn't collect names for any kind of membership, and sets up no organization. He simply moves on to another city.

But everywhere he goes, there are thousands who are asking the question, *"Is God bigger than I was told?"* Without any structure, organization, church of any kind or leadership in general, humanity still comes to hear what the ancients knew. It rings true, and many are changed forever, for the secret is huge: **You are God, and the power within you is absolute!** You can drop the cultural victimization that has been told to you for decades! You can work outside of the box your family told you that you were stuck within. You can co-create your own destiny with others doing the same thing, and yes, eventually you may even have peace on Earth instead of the doom that so many wish to convince you of.

In the meantime, the earth isn't going to crash into another planet. The crust isn't going to turn upside-down. The sun isn't going to explode. The poles may again change polarity, but never will they flip over, killing every living thing on Earth. There won't be days and days of darkness, unless you interpret that as the electricity going out! [Kryon humor] The water won't rise to cover the continents. Does any of this information sound like it should accompany the greatest Human consciousness ever seen on the planet? *Why would Gaia align with you to then destroy you?* You will practice reasonable population restraint—yes—it's already happening!—and the energy you need is available from the heat of the earth to the tides that hit the coast lines, all for free! This is the shift, and this is what you are learning about as you read this book. It's time to use what we call "spiritual common sense."

In the Christian Bible, Revelations—not the last chapter written, but the last one presented—spoke of the number of the Anti-Christ as 666. It was frightening, and even warned of this number being marked on the forehead of a Human. Now we reveal that prophesy, for it is one that is a warning not to read anything but the official doctrine. Don't stray, or you will be marked. Don't think out of the box of the leadership, or you will have your soul captured.

Why the forehead? Ask an East Indian, or Buddhist, or Hindu. Often called the *Bindi* by the Hindu, it is commonly seen as the "gateway to higher consciousness" or "the third eye." It lines up with the pineal gland in the brain, called *"the seat of the Human soul."* Why is a historical sacred mark on the forehead suddenly "The Mark of the Evil Beast? Are you beginning to understand the warning?

Now you are reading the secrets, and guess what they contain? I'll show you:

Start adding up the three DNA groups of four.

Group One contains layers 1, 2, and 3

This adds up to 6

Group Two contains layers 4, 5, and 6

This adds up to 15 or 6

Group Three contains layers 7, 8, and 9

This adds up to 24 or 6

666!

Group four contains 10, 11, and 12.

This adds up to 33. This is a master number in numerology and (as Kryon mentioned) stands for *God Almighty!*

Therefore the secret of the 666 is that **YOU ARE GOD!** The power and divinity is inside the DNA! It's the energy of the 33... *God Almighty.*

This revelation will be startling to many and revealing to the rest. What are you going to do with this information? This is where your intuition starts to work, for truth is truth, and old soul, you have lived long enough to remember—if you wish to.

This information is given in love, so that humanity may move forward into a place that enhances the planet, creates a higher vibration, allows for a peaceful countenance, and eliminates drama and fear, all the while respecting every ascended master who ever lived. For they all gave this message, all of them.

Live Channelling

Activating the DNA Field

Channelled in Boulder Colorado

Chapter Twelve

Chapter Twelve
Activating the DNA Field
kryon in Boulder, Colorado

Greetings, dear ones, I am Kryon of magnetic service. This message is going to be about DNA. Since the year 2003 I have been slowly giving my partner the messages about the energies which are in DNA. Way beyond that which you call the chemistry of the double helix, yet very much integrated with it, I began to discuss the attributes with him. In the process I told him there would be twelve energies of discussion that would be put into a three dimensional teaching involving layers so that multidimensional DNA could be taught. It was for one reason only: so that you would see the magnificence of a sacred system hiding within the chemistry of the Human body.

It is not given so that you would worship it; not given so you would do anything remarkable with the pieces and parts of it, but that you would begin to understand the overview of it. DNA goes far beyond the chemistry, and it actually represents pieces and parts of the God-self, and the Higher-Self. I enumerated these layer energies slowly and told him to create still another book. That was in 2003. But it didn't become a book back then even though he thought it would. The publisher was ready but there was no book. I gave him all the names and all the energies and still there was no book. At each place he tried to finish, he could not. Now he knows why. It is because the numerological attributes and energies around this publication had to be a *three*. There were also things to come which he had to wait for.

Back then, he was creating Kryon Book Ten and then subsequently he created Kryon Book Eleven. However, this is the year of the *three*—2010. He is now creating Kryon Book Twelve, and you are reading and hearing it. That is a *three* [the 12]. Today is one, seventeen, 2010 date [January 17, 2010—1+1+7+2+1]. That is a three. And so we have 3 threes, and a trio of the same value exists [3+3+3]. This creates a nine which is completion of Kryon Book Twelve! So on this *three* day of the *three* year, we give you the channelling in the book of the *three* [Kryon Book Twelve] about DNA.

As we speak at this moment to you, in this first channelling of 2010, we say to my partner that this communication will go in the book. He is to transcribe this channelling. The things which we do not give you at this point in time through verbalization, we will also add unto when he transcribes it. He will *fill in the cracks* for those who are going to read this instead of those who will listen, who are here. For the listening audience is going to receive an energy around what we discuss that the reader cannot have. For in this room we bring in the entourage of Kryon, who have been waiting for this moment for some time.

We would like to discuss the final leg of teaching, that which is the final attribute of DNA. It is the one we have not given succinctly and that is why, my partner, I want you to go slow so that in this place there will be developed a sense of knowing, a mist that envelopes all of you for a reason I will tell you in a moment. Truth is revealed in these moments, which is profound.

DNA

By this point I have given you all twelve layer energies of DNA. I have given you the attributes of the energies within the layers, the numerological values of those attributes, and the combinations of layers within the layers… all so that you would have an appreciation for what is inside you. Yet, there is a missing piece, and I'm going to give it to you now.

Let us review only one item: Let us talk about what DNA is, for *DNA in the Human body* is one of the only places within all life forms that has the attributes of quantum atomic structure. It is the only one where you're going to find a 3D existence with a quantum overlay. Not all DNA has this. The Human chemistry of DNA shouts to the scientists that there is a puzzle. Three billion chemicals strong, each strand or loop of DNA has within it an enormous complexity. It's a complexity, however, that only shows itself within the three percent. Of the chemistry of three billions parts, only three percent is in your seeable dimensionality [The protein encoded parts].

To the scientist, ninety percent of the chemistry in DNA seems to be inoperative. There are no codes, and there are no genes being produced. It simply lays there and does nothing. That is how science sees it in their 3D bias, looking for what it does in a linear structure.

The reason it does not work for them is because they're observing a *multidimensional event*. However, science is not looking for *an event*. They are looking for chemistry... codes and reactions. There's one other place you'll find an event like this, and I would love to take you there. If I take you into the atom that is the very central structure of your reality, there is an event there as well. Let me take you into what you would call a layman's view of atomic structure, and the center of one atom.

You will be overwhelmed with the light there, for there is light created. That is to say, there are elements of photons there. The nucleus is obvious. It vibrates. The electrons which surround it, which we have called the electron haze, are at an vast atomic distance from the center. This is why, perhaps, you've heard physicists say something you might not really understand: that most all matter is made of space... seeming nothingness.

Chapter Twelve

Stand with me at the center of the atom with the nucleus literally surrounding you, and you can barely see the electrons far away. They're seemingly miles away. In atomic structure, the relationships of the parts and the sizes makes it really hard to understand how much space there is between the orbiting electrons and yet there is something profound about it all, with trillions and trillions of atoms all patched together: Not one electron invades the space of the other atoms. They're all whizzing around the nucleus, but they're all respecting each others' space.

One electron does not interfere with the other. They don't overlap. This ought to tell you, Human being, that there has to be something in the space that holds it all together. That's the quantumness of atomic structure, or the multidimensional attributes.

There are pieces and parts of the nucleus which you have not identified yet, which temper the space [pattern it]. So you might say that it appears to be only space to you, but it is an multidimensional soup, filled with instructions and energy. So atomic structure is multidimensional, and part of that "multi" is 3D. Yet it has quantum attributes at the same time. Most multidimensional energies are like this, including light, gravity, and magnetism. They all show in 3D, yet their main structure is multidimensional and vastly larger than you would believe.

The DNA molecule is the same. I want to tell you about DNA in a way I have not discussed before. The multidimensional attributes of DNA are not touchable in 3D. Imagine for a moment you're driving in a car and above you in the clouds, it's raining. Perhaps it's not raining where you are; perhaps it's raining just a little bit further on. Perhaps it's not really raining at all, but just misting. In that mist, there's a rainbow. Isn't it interesting about the rainbow? The rainbow really does not have a tangible place, does it? For as you're driving in the car, notice it moves with you.

So it's about your perception of something that is very real. The colors are really there. They're beautiful! Optically, it is simply the prism effect with water. Sunlight hits the water mist and for your pleasure, the spectrum of the colors appear in the sky. The yellows, the purples, and the reds, the blues… they're all there in the bands that you've seen of colors since you were a child. But isn't it interesting that there is no substance to it at all? You can't touch it, and there is no end of the rainbow, for it always moves as you travel along. (It's actually a circle!)

Isn't it interesting that the faster you travel, the faster it moves? If you stop, it stops. It's the same for someone else traveling and stopping. Each Human has their own perception of the same rainbow depending on how fast they are going. So it's relational and perceptual, and unique… yet so real.

Now, let us look at the rainbow and make comparisons to DNA. There are bands of colors in the rainbow and if I started to teach you lessons about color, we could therefore talk about the yellows and the purples and the blues and the greens and the reds that are there. We could give a dissertation on each, what they mean, why they're there, and the prism affect that they represent. Yet you also know that none of you, not one Human, can ever just *go get the purple!* The rainbow is always complete, yet it appears as segments.

DNA is exactly like that. We have given you the layer energies, like the colors you see in a rainbow in a three-dimensional way, and because of the structure of the teaching, many who read this book will say, *"I want to work with this layer or that layer,"* but you can't. DNA is interactive and complete all the time. Like the rainbow, it can't be segmented, even if you study it in a segmented form.

I want to give you the attribute of DNA that is more beautiful than anything we've ever talked about. I've waited for the *three* [the date of the channelling]. Let us discuss the *three* so you'll know why this particular number is so important. In numerological terms, the three represents a catalytic energy. *Action* is what the energy is. In DNA, the third layer is the action layer, which *creates*. Therefore it is the *activation and ascension layer*. It is the energy of the *three*, the catalyst. A catalytic energy is an energy that arrives and changes what is around it, yet the catalyst remains the same. That's what's happening in the room right now... the *Third Language* is here.

In order to teach this properly, I wish to walk this assemblage in the room through a metaphor. Metaphors are designed to let you understand or think about things that are often complex or beyond your perception. So they become 3D's examples like the metaphor of the rainbow, which you can see but you can't touch.

I want you to imagine DNA for a moment as a single source. Imagine one hundred trillion pieces of chemistry all singing the same song. Nothing else in the body does that. You may call it your soul energy if you wish, for it is. DNA is identical from the top of your head to the tip of your toe. Science can identify you no matter where your DNA comes from within the body. It's all identical. You might say that it is all singing in unison for there is an *awareness* of DNA. It's an awareness at the soul level of the entire structure of the Human being. We have told you before of a phenomenon of Human energy that reaches out past the Human body, up to eight meters [26 feet]. You have a name for it, and coincidentally, it is a Hebrew name, just like the names of DNA.

It's called the Merkabah. Hebrew history has created the name, which was created through the ascension of Elijah, as seen by Elisha. Loosely translated it would mean "to ride." Elisha saw the divine field of his master activated at one hundred percent, and the metaphoric image of three horses carrying this vehicle he was then riding within.

So what does it got to do with DNA? There is an *awareness,* a multidimensional mist that all of you have, which expands out to eight meters. This *mist* is created by the *DNA event,* a hundred trillion of pieces of it working as one. This creates the Human Merkaba. So, what is it that rides upon this energy? I will tell you. It is the pieces and parts of DNA that I've given you that are multidimensional. They ride within it like the colors of the rainbow. They are the vehicle which your soul energy rides within.

Let me again list these parts for you. They are your championship over duality. They are your activation of the ascension that is yours as you walk Planet Earth. They are your angelic name. They are your divinity. They are your Higher-Self. They are your God part. They are your Akashic record. They're your Lemurian history. They are your healing ability and belief (faith) abilities... all in that rainbow-like, interdimensional mist.

Imagine this multidimensional mist to be a rainbow. Imagine the colors. Imagine that you push this rainbow mist from place to place as you walk. However, also understand for a moment that you come in with a default setting of Human consciousness where you've never talked to your cells, where you have no idea about a Merkaba, and where you have no idea about DNA. The rainbow is there, but it has no direction or "boss" who it telling it what colors to *show.*

The default setting at birth is a DNA operating at thirty percent. But it still has within it the full range of the rainbow. However, it's only showing the colors at thirty percent brightness. [The vibration that you have created by choice on the planet]. Now imagine for a minute in your mind, the masters who have walked this planet... think of all of them. These are the ones who projected peace and joy wherever they walked, who the children followed, and who the animals saw and wanted to be with. Imagine a master who could touch a Human and heal him! These masters had the same DNA you have, but with *the colors* at one hundred percent.

Activation of DNA

With awareness comes *activation of DNA* and that is what I wish to speak to you about now. Hold this misty metaphor in place. Everywhere you walk, Human being, the mastery that is you, rides in the vehicle of the Merkaba, created by your personal multidimensional DNA. Like the single atom of your creation, you have "space" around you which is filled with patterning, but which can't be seen in 3D. It's because the DNA is a multidimensional, spiritual *event*. In the future, you are going to hear more about the activation of DNA than about any other process in metaphysics. So let us look at the history of the way this has been developed.

The Four Steps of Human Evolvement in Healing

You have healing attributes of Humanity all around you. The first one is obvious, and that is what you call the allopathic. It is a straight line, *cause and effect,* process, one chemical creating an effect through chemistry on another. It's very 3D. It's the one modern science trusts the most, since it sits firmly in the 3D reality that you can see. It represents modern medicine and the drugs that are able to help heal your body.

The second one would be a *signal to the body* for chemical change, without the allopathic attributes. That is homeopathy. A tiny tincture, that is to say a very small chemical… too small to create a chemical reaction… one in a million parts, creates a *signal* to the body to create a biological reaction within itself. You might say it's a signal for the body to do the work, not an external drug or chemical to do the work. So what is it? It's information!

According to those in science who see only chemical reactions as their reality, a homeopathic reaction is an impossibility. This is because they are not giving credibility to *body awareness of information*. DNA is aware. All of it together has a singularity of awareness that is multidimensional, sitting in a quantum state, and it all *knows* together. It is aware! It is more aware than you are of what you need, both spiritually and biologically. Yet all of that knowledge is there for you to tap into, but not at the default setting.

Working with the twelve meridians of the body is within this second area of evolvement, for needles and pressure, and even colors and hot and cold applications to certain energy meridians can also create signals to the body to heal itself. But that is still very 3D.

The third step of development would be *pure energy work*. That would be energy which sends messages to the body to heal itself. This would include energy from those healers who can help balance and heal without touching you. It is the beginning of actually addressing the Human Merkabah [DNA field], creating healing and balancing energies within the Human body using new information that has only been available for 25 years on the planet. That's quite an advancement, isn't it? But nothing compares to number four.

Number four in numerological terms is a very, very earth-based number. It is actually a profound Gaia number. It's very practical at a down-to-earth level. It's the most intuitive practical energy you can have (Page 86). *It's you addressing DNA at the core level through Human consciousness.* In this new energy you have the ability to speak to DNA all together. When you do, you're speaking to the *aware mist*. You are addressing the rainbow of your own DNA field, and it is aware of YOU!

Yet you will tend to linearize it all. You may want to work on something you heard about in layer nine in this book. So what does the Human Being do? He tries to *grab the purple,* in the rainbow and use it. *"Dear layer nine…"*

[Kryon laughter]

There's no punishment for that. It's *3D normal!* So this book is about showing you how to consider your communication with your DNA in a non-linear fashion. When you address your DNA, you are addressing the mist and the entire rainbow will decide what layer it's going to activate. Not you. Again, the DNA field is a *knowing*. It is not linear or static.

Let us speak of activating DNA...

Again, there is really no such thing as selecting a layer of DNA to activate. However, if you were to consider which layer has the most influence over activation, it is Layer Three. But this is an internal event that activates this layer, not the Human Being. The Human can only address the entire DNA field. Again, the metaphor: The Human might say, *"I wish to make the purple in the rainbow brighter"* The rainbow then activates the energy of all the colors to accomplish that. It knows what to do, how to make the purple brighter, and is complete in its *knowing*. You can't really address any color individually... only the entire rainbow. This is one of the reasons sound and color work for activation... did you consider this? The body selects what it needs from the variety that the healer provides.

In this metaphor, it's hard for the Human Being to look at the rainbow, knowing they wish to use the purple layer, but trying to understand that only the entire rainbow can do that. It's elusive too, since it's only a mist, and not something you can hold in your hand like a chemical.

I will give you a truth now. You can call it whatever you wish. Blessed is the Human beings who activates their DNA through any process. Through meditation, through music, through water... through design processes, through energy work, through energetic counseling, through touching or non touching. *Since you're working with an innate aware system* of DNA, it knows what you are trying to do! It has always been this way. You can activate DNA and there are many ways, many processes, many substances. But the truth is that the awareness of DNA will decide what layer or what energy it's going to work with. Not you.

"I get it, Kryon, but still I've had many healers tell me that they're going to activate one layer or anther to help me with my issues. Are they incorrect?"

No. What the healer is doing is to help you communicate to the whole of DNA, to get to the energy that you discern in 3D as one part or another. So this 3D verbiage is with you and is fine. But the truth is that DNA is aware, and within its intelligence it will take what you give it and decide what to activate. It knows far better than anyone, what you need. That's what you have to know. You may wish to compartmentalize it all because the teaching is compartmentalized, but you can't.

There will be many of you who will say, *"Well, I have this or that healer who is activating this layer or that layer…"* You might say, *"I have one healer who believes that they can increase the vibration of DNA."* I say, congratulations for that awareness! For what they are doing is real, but what is going on is a simple linear verbiage over a non-linear process. Honor the process, and don't try to put it in a box of layers or energies, as you do almost all the other processes in your life. This one is different! These healers will get results, for the intent to work with the mist of the field, creates the needed communications that DNA will then act upon. The healer may actually believe they are activating a specific layer, but DNA knows, and reacts accordingly without judgment or comment.

Blessed is that healer who sees the scale of DNA activation. There will be those with second sight who can see the percentage level of DNA activation... of a Human being who sits in front of them. What are they doing? They're actually *measuring the mist*, and while they measure the mist, the DNA is aware of the measurement going on!

I want to tell you something you didn't expect. I want you to listen because this is the end of this lesson. Imagine a mist is around you, the Merkaba; all that is you; your Akashic record; all of your lifetimes, they're in the mist. Your Higher-Self is in the mist, and that which is divine is in the mist. The healing layer is in the mist. All this is waiting... waiting. In comes the healer who wants to measure it. Can you imagine what the "aware mist" of the DNA field *thinks* at that moment? I want to tell you the energy of that aware mist at that moment: *"Hallelujah! A Human is helping my soul partner to pay attention!"* The field knows that the Human wouldn't have it measured unless they were aware of it, and ready to do something... perhaps even communicate!

One hundred trillion of pieces of awareness are awaiting for you to talk to them all at once! Blessed are those who are in an activation mode, for it means they're going to begin a process of cellular communication in their own way.

Finally, we give you this: At the center of atomic structure, there are magnetic fields. The physicists know about them and they call them interdimensional fields, but they also know the fields have to do with the quantumness of magnetics (at very small energies). Every piece of DNA also has a mini-magnetic field. Together, the DNA molecule's fields overlap to create a sonority of love. The creator is in you, and this energy is not all that new, Lemurian.

The invitation to communicate is upon you, and so this is given in this publication so that you will understand when some say they're going to activate this and that layer. It is very appropriate within your 3D understanding to visualize this, for you're simply getting communications of what you want to work on. DNA does the rest.

And so it is.

Kryon

Riga, Latvia - March 2009

Live Channelling

Beginning to Activate The Specific Energies of DNA

Channelled in Riga, Latvia
March 2009

Chapter Thirteen

Chapter Thirteen
Beginning to Activate the Specific Energies of DNA
kryon in Riga, Latvia

Greetings, dear ones, I am Kryon of Magnetic Service. It is again we say that the energy that is presented to this place is pure. Again we say there is no agenda or manipulation. There's no plan here to capture your will; only to open your heart.

Human Beings search for the Creator, and they do so almost from the moment they're born because there's something missing. I'll call it the music, the song that they have heard for eons on the other side of the veil... my side. It's such a shock to be born into this planet in the Human body, to suddenly have the cellular structure presented to you yet again. It's a shock to the system that is the soul. The thing that is most obvious from the one side to the other is the feeling that you are suddenly alone. Then you spend the rest of your life looking for the missing piece.

Some of you have discovered, perhaps in your houses of worship, what that feels like to kneel and call upon the power of God... and to have it wash over you. If that's the case, you just want to sit there, don't you? It's beautiful. Some of you, in your meditations, have the same thing. Beautiful, it is! For you can sit and be in touch, you think, for a little while. It's something physical you can do to make that connection that is missing.

In those brief moments, all is well. In those moments you feel you are creating a time where you touch the face of God, just a little bit. And when you do, you hear the music and all is well. You spend a lot of time trying to recreate something that is missing... missing from birth.

Chapter Thirteen

Humanity has always done this. I know why, for the music never stops for me. I will tell you, as Kryon, I cannot imagine being without it. The celestial choirs that sing *in light* on my side of the veil are what we are all used to, because we are eternal and belong here. You have spent eternity in both directions, listening to it, too, yet it's completely disconnected when you arrive on Earth. Frustrating it must be to the Human Being, that it went away... that it's not there.

So we begin the teaching today, and we begin by saying this: There is a new energy on this planet, and this energy seeks to now activate pieces and portions of your soul energy that we call DNA. It is that part of your body that is both in 3D and multi-D, where the Higher-Self resides. As you activate the energy from the tools that are new, you also activate the realignment of the magnetic grid, the establishment of the Crystalline Grid, the vibratory cycle of Gaia, which the Mayans and ancients predicted... all of these things are in play. Then you stir in the energy of the Harmonic Convergence, all of which gives you permission to begin a process that is a sacred graduation, which is mastery.

These are the mastery tools— the allowance for you to understand a multidimensional state, which is within you. In the process of this discovery, some of you will begin to hear the music again. I promise. It won't be something that you have to sit and meditate to accomplish because DNA, when activated [energized], stays that way [stays changed]. You won't be turning it on and turning it off. It is a complex puzzle, this, and I challenged my partner this night to be careful and go slow so that it is directed properly, explained succinctly, and communicated with love.

What follows is going to be transcribed into a book, so these are instructions that my partner [Lee] is hearing for the first time. For the DNA book is in progress and this will be a chapter of it, for this is the practical part of something that heretofore has only been academic... the 12 layers of DNA, how they are identified, what their names are, what their purpose is. But now there is even more to this story, and tonight I choose to give this information to you here in this great place [Riga Arena], this place of many languages [The Baltics], this place of recent freedom.

There's a reason why we reveal it to you here, and part of that reason is because this is not a Western message! It is a universal message for the entire family, and they sit before me now, listening to many languages [English, Latvian, Russian, Lithuanian, Estonian], but feeling only one language— the third language, of love.

The Teaching Begins

There have been questions by those who attend channellings like this one, for they ask my partner, *"What's all of this about DNA? Why does Kryon always talk about DNA? Why doesn't he talk about angelic things, esoteric things? Why doesn't he tell us what we came for? But instead, he just wants to talk about chemistry."* If you are one who has said that, dear one, you still don't understand the big issue that has been revealed: Within your cellular structure are the tools of mastery and they're not invisible. They are part of the cellular structure, which is being studied in an entirely different way by science today. The spiritual secrets are in the DNA! The instruction sets for mastery are in the DNA! Therefore, an esoteric study of your DNA becomes a study of mastery, the God within, and an entry to all that is.

You have a hundred of trillion copies of the double helix in your body, and I've given you the names of the energies within them. Now we're going to speak of some specific ones, a set of them. However, before we speak of the details we have to once again give you some of the scenarios that you must understand.

Human Beings want lists! You wish to compartmentalize everything so you can linearize in your mind, how things work. It helps you to learn, since all of you are used to linear thinking. You are also used to singular purpose mechanical and chemical functions. Even the most complex chemical functions work the same way every time. You find out what does what, then you apply it to a list of attributes that occur in a certain order over time. Eventually you become a chemist who knows these linear processes.

Multidimensional DNA does not work in a linear fashion. Therefore, you cannot compartmentalize it in its function. Here's what I mean: Within the most complex machines on the planet, ones that may have tens of thousands of parts, the parts always do the same thing. You may have something a thousand times more complicated than a fine watch, but the springs and the gears are always springs and gears. They do the same thing over and over in a complex way.

Your most sophisticated electronics do the same thing. They provide millions of identical processes over and over, and the electron paths always do the same switching… a singular process done fast with linear complexity. Not DNA. It is not a machine. You must start to think of DNA as being totally and completely interactive with itself. When one part changes, the part next to it also changes. You cannot then identify a singular purpose of specific DNA parts, always doing the same thing. Think of that complex clock we mentioned. What if it were interdimensional, and the spring could suddenly become a gear, and the gear could change shape and size as needed? Think of this— a part that is no longer needed, simply vanishes and if something is needed that is not there, it appears! That would, indeed, be complex, would it not? In addition, this quantum clock decides on its own to change the time frame it was designed to work with. Strange? That is quantum DNA.

Some time ago, we gave the explanation of the 12 layers of DNA as they exist in an academic way. We were careful, however, not to tell you what each layer did all the time. We only gave you their purpose and their energy or their storage attributes. You'll see that in a moment. But as they work together, each one morphs; that is to say, it changes depending upon what the other one does. So you might say you have an engine where all of the parts continue to shift and change depending upon what the engine needs. Complex, it is, and quantum, it is.

In a quantum state, there is no time. In a quantum state there is no actual place where anything is. For quantum mechanics dictates that any matter, if you want to call it that, or any energy, if you want to call it that, is everywhere together, all as one. Imagine something so complex! Now imagine that it's duplicated hundreds of millions of times in your body.

There's something else. Even before we begin to list these parts that are the new tool set, there's something else.

[Pause]

The reason my partner is hesitating is because I am showing it to him, and there are no words. There is no science, yet. It's new to him. In order for DNA to work properly, in order for it to do all of the things that we speak of esoterically and quantumly, 100 trillion pieces of DNA must all know something at the same time! There has to be a communication that takes place in the microscopic DNA of your toenail at the same time as the longest hair on your head. They both have to know about it instantly. Then those trillions of pieces must agree, must have one energy absorption of consciousness. This all must happen in a 3D construct— that is, within your reality. There is no word in science for this process unless you consider the one created for a description of photons called "entanglement."

There is instead, "a confluence of energy." Confluence in English truly means a melding of energies together, so that they become something else, a oneness. Science doesn't see it within DNA yet, but at some level they know it must exist. For how else can the Human body do what it does?

Cellular structure is specific and unique. Like the linear machine, it is specialized, and you have heard of stem cells carrying that specificity. But DNA is identical all over the body. It's not specific. You don't have toenail DNA or hair DNA or heart DNA. You've just got your own unique DNA. Trillions of copies of the same Human quantum blueprint must talk to each other instantly or you would cease to exist. How do they do it? No name in science truly has been given to the process of communication between DNA loops, but it will. It's a quantumness within the "soup" of magnetics.

Let us study now how the communication is accomplished, and the puzzle starts to come together. For the magnetic master is before you. Many have thought for years that the magnetic attribute of Kryon has been about the grid. This is only partially true, but it's really about the magnetics of DNA. For as my partner indicated in the lecture series earlier today, there's something you should know about DNA, and that is, it has magnetic components. Each loop of DNA has a magnetic field that overlaps the loop next to it, which overlaps the loop next to it. Trillions of overlaps equals one consciousness. This then represents a magnetic imprint, which Humans carries around with them.

Magnetics is an interdimensional energy, a quantum energy, and this imprint creates the Human aura. An aura is *not a magnetic field* and you will not be able to see an aura with magnetic equipment. An aura is the result of a confluence of DNA communication within the Human body, a quantum imprint, a melding of energy to create a quantum field not measurable by anything on the planet, yet.

I tell you these things because you have to understand that what happens in DNA, happens all at once within every energy layer of it. Think of the coordination, the puzzle. If you're going to have some kind of esoteric activation that is new within your DNA, think of what must take place! Trillions of parts all receive it at once. What does that feel like?

Reflect for a Moment

I have been channelling now for a few minutes. However, in a crowd this size there are many spiritual paths. Some are beginning to realize that this is a real communication and some will never realize it. There will be those who understand what is to come and there will be those who don't. That is why it is being transcribed the way it is in real time so you can revisit the information when you need to. But nothing will ever give you again what is happening right now; you and I are in a quantum state, dear family, right at this moment. Take advantage of it. For in this place, there are quantum energies walking between the aisles. Some of them are touching you right now and you can feel them.

Specifics

Let us talk about the DNA. This now, my partner, is critical. Get the numbers correct; get the purposes correct, and teach this in a fashion that allows understanding. [These were instructions for Lee.]

In this new energy, there are five layers, or energies that are being activated to a point where great gifts will be yours. But I would be wrong to say that it was happening in a linear way, for one of those five is always being activated. So really, there are four, plus one! Already we're in trouble because we are linearizing quantum energies, trying to count them. We're going to do it anyway, because there is no way I can communicate quantum things to you unless I reduce them to a three-dimensional construct for teaching. So I will.

Five quantum energies are involved in the gifts of this age. Specifically, five. Yet you might say we can't ignore the main one, for it's the main 3D layer. If you said that, you'd be right. That would really make six energies, but I'm still going to stick to a concept of five, since DNA Layer One is always affected by anything that happens within the others. Layer one has been identified at the 3D double helix... the linear chemistry that you can see under a microscope. Although the parts do not change, how they interact is programmed by the other 11. Therefore, number one is always in shift. It represents the chemistry; therefore, it represents what you see and feel.

In numerology, five is the number of change, and this is proper for us to place that energy upon this lesson. The five energies or identities of the layers are *number two*, *number six*, the combination of *layers seven and eight*, and *number nine*. These represent five energies involving six layers. I will explain as we go, and will list them in review. This is esoteric teaching and it is complex. It is beautiful, new, and it starts to explain some things for you, concepts you need to know. It provides an explanation of the beginning gifts of the shift, and provides attributes you can do that you've not been able to do before. It starts to explain the unexplainable.

Already I've given you the overall numerology of this lesson. So already, we're into the energy of numerology. If you doubt that we are dealing with the energy of five, then do this: Take layer two, layer six, layer seven, eight and nine and add them together, you are going to get an energy in numerology, of five [2 + 6 + 7 + 8 + 9 =32]. The 32, when reduced to one number, is *five*. So again, five means universal change. Get ready for change!

Now, I give what my partner calls the disclaimer. [Kryon smile] None of these things are going to happen to you if you don't want them to. These are gifts for Lightworkers. If Lightworkers want to open the door, if they want to push upon the quantum parts of themselves through a consciousness of choice, then change will take place. If they don't, nothing whatsoever will happen. Your reality on this planet is driven by Human consciousness choice.

For those who don't understand this concept, I again repeat to you the very famous Hardy's Paradox. It states (in your own science) that in a quantum state, photons that are observed change their attributes. This truly is no different, for your intent is the observation to create change for your quantum DNA. Without intent, nothing happens.

DNA is very much like the most advanced jet airplane ever made. However, the energy on the planet earth, which literally talks to your DNA through the magnetic field, has always been low. Compared to the energy of mastery, compared to the prophets, your DNA has always been of low energy. Therefore, that wonderful jet plane has simply taxied up and down the runway for all of humanity, for all time, all through history. Now it is changing, and you are ready to fly. That's the message, and always has been. Let us itemize what is different.

Teachings of DNA - Quantum Lessons for You and Your Past

Layer two of DNA has been described as the Human Being's *life lesson*. When you compartmentalize DNA into energies, then put it in 3D for teaching, it's easier to understand. It's important that you understand what a life lesson is, and why it's in your DNA. You come into this planet with it, and it is something that has been constructed for you due to the Akashic Record. That is to say, your current life lesson is connected to your past lives.

Therefore, because it relates to past lives, your personal Akashic DNA record, it is linked to layers seven and eight [the Lemurian layers of your DNA]. Already, we're interacting, are we not? No layer works alone, but let us return to number two.

For those among you might say, "Life lesson? Sounds like karma," it isn't. It has nothing to do with karma. Instead, it is an energy placed upon you that some have called purpose. What is it that drives you? Some of you come in and the cells sing with a message that says, *"I don't ever want to be alone."* What do you think your life lesson is? To be alone! Doesn't seem fair, does it? But that's the energy of the life lesson.

Whatever your past lives have put together as a puzzle representing things you have not yet accomplished actually becomes your life lesson. There are many of you in this room who have a very similar life lesson and when I tell you what it is, you'll know I'm right. For generations, the Baltics have had a reality of life that was defined for you. You are this, you are that. You will think this way, you will think that way, and many of you know of what I speak. Therefore, there are so many life lessons in this room that are the same: Never let anyone define you as a society again. Let the freedom and the independence that is yours countrywide, be reflected in your thinking. This is not karma, but that is personal and reflects what you do with others. This, instead, is your life's lesson.

Do you know what that does for you as a group? It lets your energy soar! Don't be surprised if there's an awakening here in this place, which is spiritual based. That's the way it works. That's layer two, and it's being changed. It's being modified. It's being altered. So you might say, "Well, how it is being altered? The life lesson must be the life lesson. How can it be altered if it's based on what actually happened?" I will answer it this way: It's being changed in its ability to be manifested. So the lesson may be the same, but the result of the lesson is enhanced greatly. Stay with me.

The layer that is always involved in everything is *number six*. Do you know the numerology on six? Look it up, for we told you this: Number six represents the energy of the Higher-Self.

If you haven't heard this before, you need to hear it now: Where is the Higher-Self? Well now, based upon the actual words, it has to be higher, right? But it's not. The only part that is higher is your perception of its vibration. It's in a place that makes you want to worship it. Did you realize that so many of you go into the houses of worship to connect with yourself? So you kneel and you hear the music and you pray to yourself!

I am not being blasphemous. I am giving you core information on the way Humans work, and your perception of God. The Higher-Self, that grand angel, the golden one, that piece and part that is connected to the other side of the veil, which is your core soul, always seems to look like God to you. Yet it's part of you... and it's in your DNA! You'd like to be connected to it all the time, wouldn't you? Well, until this new energy developed, it required that you work for it, because it was not within the normal reach of Human consciousness. Now it is!

What about a 100% continuous connection to the Higher-Self? *"Well, now Kryon, if I had that, I would be like a master."* Exactly! And what keeps you from thinking such a thing is possible, dear Human Being? Is it a little too grand for you? It is not too grand! It's the feeling of purpose of life, that there is a sacred reason you are here. Oh, the challenges come and go, but that's just life. The love of God is always in your heart, always connected. The energy of the shift is starting to change the ability for you to have this, and it is now becoming far easier to achieve this mastery.

Layer six has to be involved in every single tool set I will ever describe to you. So the intellectuals who want to place things into 3D will be counting up the layers for numerological reasons and saying, *"Well, if it's always involved all the time, do you count it or do you not count it?"* I'll tell you it doesn't really matter. Because it continues to change with every layer it's working with. Why try to catch this elusive energy that is quantum? Do you really need to understand all of it in order to use it? So, layer two is the life lesson and layer six is the Higher-Self. They are both involved in this tool set. Now we come to the two that are critical, and which we mentioned earlier.

The Two Most Important Ones

The two energies (layers) of DNA that I'm going to give you now are the most important Human DNA energies on the planet. They will always be the most important, since they drive the engine of karmic purpose. They are responsible for life lessons and they relate to layer six, the Higher-Self. Listen: There are only two DNA energy groups that are pairs. They are layer four and five, which we will not discuss today, and layers seven and eight, which we will.

Like four and five, seven and eight are married together. You might say, "Well, why is that? Why didn't you just call them one layer?" It's for emphasis and for numerologists. [Kryon smile] You needed to hear how big the energy is in order to understand, for layers seven and eight are what we have called *The Lemurian layers*. They are *the creation layers*, which are your Akashic Record. That is to say, the record of every single lifetime you've ever had on the earth, everything you've ever done, all the accomplishments, all the talents that you have learned, and the spiritual jar of knowledge that you have filled up along the way.

We have given you channel after channel explaining this energy to you. I'll say it again: In your DNA there is spiritual knowledge that you have learned for eons. How many past lives do you think you have been in, old soul? You sit here with no proof that you have been in any past lives, but you intuitively know, don't you? Oh, you do! I know who's here, and you can't deny it.

Along the way, you've learned what you know now. Along the way, you've picked up the pieces and parts of spiritual purpose and learning. You also have made all the mistakes you needed to make, and it fills up the knowledge in the spiritual jar in your DNA. Then, in you come, this time around, as the Human Being who sits in the chair in this room. So I ask you, have you opened that jar yet, or are you going to make all the mistakes again? You need to know this: In that spiritual jar, in those layers of DNA called the Akashic Record, are lifetimes of knowledge. Should you choose to open this quantum jar with intent, out will come shamanic energy! You'll be so much wiser for doing that, dear one.

I've just described one of the attributes of the first gift. The ability to have spiritual wisdom immediately, so that you will know your purpose, your life lesson, feel the love of God in your life, and start a process of *connecting*. Just by opening the spiritual jar, you get this. That's a big one, isn't it? It will lead you into what you need to know, and it will lead you into synchronicities for the manifestation of what you're here for. Those who would open their spiritual jars will also have knowledge of what the courses are in the new energy that they need to take, for many are new. That will help enhance what else I'm going to tell you, which is quantum. It all fits together, eventually.

Now, let's discuss further the interaction of layers seven and eight with layer six, the Higher-Self. What is it you think is important about past lives? I'll give you two attributes to discuss that are not new, but they relate to this discussion.

In a quantum way, there is no such thing as a past life. Put that in your notes, for your DNA and the Akashic Record is a quantum state. Therefore, you cannot use a time reference like the word "past." *"What then, is a past life, Kryon? Didn't I live it in the past?"* Oh, in 3D you might have, but the experience lives in your DNA as a current life energy. *"What does that mean, Kryon?"* I'll give this to you in a way that you can understand and find easy. What if you were a great painter 300 years ago? Did you know that talent is still there? It's still there! *"Well, Kryon, I can't even draw."* I see. So you've made up your mind and will say to me, *"No. I've tried and I can't draw!"* I say to you, *you only tried it in 3D!* What if you went into meditation and talked to your cellular structure and said, *"I would like to pull upon that lifetime in which I could paint."*

I'm going to tell you something that you're going to learn slowly: Start the process! With intent, claim that talent. You'll start to draw, poorly at first, but then something happens: Colors will start to appear in a way they didn't before. You'll know what goes with what. My partner knows this is true, for he has pulled upon his Akash in order to do some of the things I've asked him to do. It has taken years, but he found what he already had, and now you get to see it in his work. So, your Akashic Record is current, not past. Make a note of this.

Now, how does that relate to your Higher-Self? I will tell you: When you look at a past life, as you call it, let us say it was hundreds of years ago, and you're in some kind of a past life regressor's office, they may tell you that you were someone else in the past. They may give you information about who that might have been. They can do that, for it is written in your Akash, and "readable" by many. How do you feel? Do you feel connected to it or do you feel apart from it?

Well, here's what I want to tell you: The Higher-Self, number six, *was the same Higher-Self then, as it is now.* You have a connection to everything you ever did, since. Your Higher-Self was all of them! It's the Higher-Self you have now… that piece of you that which is so precious, that was the same one every single time, hundreds of lifetimes, that same Higher-Self. Do you know what that means? It means you've got a friend who was there!

That's going to play into the next teaching I want to give you. This gets complicated, and I know we're drawing this out [making it long]. But this is important that you hear. In your Akashic Record are attributes that you need, dear Human Being. This time, right now, in this new age, you need these tools. You need this mastery. You need this quantum patience. Why don't you go into the Akash and ask your Higher-Self to activate it? Your Higher-Self is going to go get it and make you a patient person. Do you believe that?

"Kryon, I can change my attributes of my personality?" That's what this is about, and that's the new tool. Oh, it goes beyond that. How would you like to be slow to anger? How would you like to have a personality that never goes into drama or fear? How many lifetimes have you had to choose from? You could go get it. It's there. It lives in your DNA. That's why it was stored there, Lemurian. Now go get it!

Mining the Akash is what we have called this. It's the new tool. Go get the pieces and parts that are you, which you lived and you deserve. What have you learned in this life? If I gave you time, if I gave you all evening, would you then start to enumerate what you've learned in this lifetime? As a woman, as a man, as a Human? And you'd say, *"Well, I'd need a lot more than one evening to write that down."* Yes you would! Now, multiply it by several hundred lives and you've got a library of experience and knowledge. You've got a large storehouse. You've got an immense amount of experience, but instead of being in the past, it's all *now*. It's all you.

Have you heard of those child prodigies who could paint like a master? They're around, you know? Don't you wonder how that occurred? Well, I just told you. For their past lives and their current life is so transparent, they come together when the Human is born, and they just pick up where they left off.

Now, some of you have done this in other ways, because what you've gone through in a past life was fearful. Perhaps you've come in with the same fear you had last time around? There are some who even say you're cursed with an energy or some sort, and you can never get above it. I will tell you that this energy is in your Akash and you brought it in... a fear that's so strong it affects your life lesson, your Higher-Self and all of your Akash.

Isn't it time to get rid of it? You can! You can reach right there into those energized layers of all of those lifetimes and pick out the one that had the hero, self-assured, courageous, loving, peaceful, and healthy. Have you noticed I didn't give you the last one yet? Now I will.

The Ninth Layer of DNA

We've defined the ninth layer of DNA as the healing layer. Now, in 3D you get very excited, and you might say, *"Well, I want that one,"* as though you could reach in and pick it up, but you can't. They're interactive, remember? I've said this before. When you sit before a bowl of soup, you can't reach in and pick out the flavor, or the tomato molecule. You can't do it with DNA either, since it's interactive.

It's the *flavor of the DNA* that you taste, that you smell. How would you like to change the flavor of who you are? How would you like to pick up the mastery that happened in another life? The tools are here for this!

Layer nine is called the healing layer, not because it heals the body— we never said that— it heals the Akash! Now we reveal some secrets: All of these layers work together, and some of them have laid dormant for all of humanity's time on the planet, ready to be activated when the earth's energy reached a certain point. That is now. I speak to a group of people tonight who know what it's like to be confined in an old energy, and suddenly have a release and an independent spirit. This is why I chose to give the message to you here first, because you can relate to the expansion of consciousness, to the joy of release, to the idea that anything is possible.

So layer nine heals the Akash. *"What does that mean, Kryon?"* It means that you have the ability to go into your own Akashic Record and sort it out. In your own way, you are healing yourself by creating a change in your DNA through the Akashic Record.

Have you come into this place with a Human body that has something wrong with it? *"Kryon, I have a pre-disposition for a disease. My parents had it and my sister had it. I might get it."* Now you're not going to believe this, so I want you to feel the purity of this message. I want you to feel the love in which it's given, for I'm giving you things that sound outlandish, that sound impossible, and that only the masters who walked the earth could do. I'm going to tell you right now, dear Human Being, that you could sort into your own Akashic Record and find the purest DNA that was ever yours, pick it up, and place upon the attributes of your current health. That's what layer nine is all about… healing the Akash.

"You mean I can heal myself?" Oh, yes! That is why you're here, and you're poised and ready for me to tell you how. It's easy and it's difficult, depending upon how much you believe in this. Is this a real communication from the other side, or is it not? If you're sitting there and saying, *"It's only a man in a chair,"* then have a nice life because your DNA is going to remain identical to what it is now. If you want to be part of the shift, you're going to have to start expanding your consciousness to include the things that are unseen.

Do you see it yet? Do you see what the differences are? You think it's chemical don't you? You think you can perform some kind of miracle and un-do a chemical predisposition. That's not it! It's *informational!* The predisposition is information within the DNA that tells you that your culture is doomed, or that you are going to die young. So... change the information!

Can you feel the love of God here? If you can, it's a good beginning. I'll tell you this: If you can *feel* this message, you can accomplish what it teaches. This healing doesn't happen all at once, but I'll give you the method. Each one of you is unique on the planet. Each one has a life lesson that is unique. Each one has a unique pattern of lives lived in the past. That means your Akashic Record is like no one else's, and your Higher-Self is also unique. This limits my ability to give you some kind of 3D generic list or way to heal yourself, since you are dealing with the specifics of "you." So instead, why not sit down in a quiet moment, with purity, and say to Spirit, *"I would like these things. I give permission to activate the energies that needed to be activated in my life, to accomplish the purposes I came for. I want to have joy in my life and to find the joy in my full Akash, for I deserve it and I've earned it. I've had positive, joyful lives, so I want to pull on that energy. I want to inform my DNA that I am a master!"*

Let say you want self-esteem. You call it out. *"Spirit... body, I need self-esteem, because I don't have much."* What happens next is the Higher-Self will go into the Akash and bring back the warrior on the battlefield... you in another life. This individual stands tall among Human Beings, and is courageous in battle. You are not asking for ego, but rather "an assurance of personality." Remember, you are asking only to dip into what you have already created and bring forth that information.

For the one who wants to write the book, or go find the orator and the author in their DNA, do it! They're there. How do I know? Because the very people who will ask are listening and reading. They are the old souls of the planet and have the fullest Akashic records... the most experience of any Humans.

For the one who wants spiritual knowledge, go find the *spiritual jar!* In there is the Shaman you were, or the teacher of spiritual things within your village. It's all there. I know this because I know you!

Review

So, there are five layer-energies that are being activated: Life lesson [layer 2], Higher-Self [layer 6], two Lemurian layers [layers 6 & 7], and the healing layer [layer 9]. These are the gifts in this age. Interdimensional, they are, and I cannot close this message at this point in time without now asking you to throw away the numbers. You're not allowed to see them as separate, as I have taught them to you. Instead, I want you to see them as a core of light that pulsates… a doorway. I want to tell you about that doorway of activation: It pulsates with the *music* of the other side of the veil.

In your meditations, if you were to push on that light with pure intent, what you would find is the *hand of the creator reaching out to you!* The reason? Because the creator wants to be found! Did you know that? Were you told that the road to find God would be difficult? Were you told that you would have to suffer, or that God doesn't really care? Nothing is further from the truth! For the creator wants to be found, because the creator *is family*. When you push upon the door, you're going to have help. It is not a task that you have to do alone. It is a cooperation energy, a partnership. I'm going to use this phrase again: *DNA is a quantum partnership.*

There's a shift happening and if you look at it only in three dimensions, it is filled with change and fear, anxiety, and uncertainty. It might be the weather or the economy or trouble with family or friends or even your job. That's the shift. However, there's another shift going on. If you can look at it in a quantum way, it's the shift we have spoken about for 21 years. Now, you are being given all the tools to handle it. All of them.

Chapter Thirteen

All of these things are wrapped up in what the teacher, Peggy, is calling the Lattice [Peggy Dubro's teaching about The Cosmic Lattice]. The Lattice represents the personal Akashic attributes of a Human Being, a quantum energy that surrounds him that can be used and pulled upon. It includes healing, a balance of life, and even mastery. The DNA is only the physical vehicle of a quantum state. A quantum state is everywhere. This is going to be tough for you to understand, Human Being, but there are really no pieces and parts of your DNA right now. They are all over the Universe! God knows who you are, oh, Quantum One. If you start activating these pieces and parts, your Lattice will change *color*, did you know that? Oh, but we'll leave it to Peggy to tell you about that. [Kryon wink]

There are new energies, and today we gave you some of the descriptions of the new tools related to this. This all started years ago, as much as 21 years. It was so in order to prepare you to sit in a place where you'll have help, now. All that my partner has been through and all the stages he has sat upon has prepared him for this moment. For he can begin the teaching and write the book about the DNA of the Human Being. Spectacular, it is.

That's what the message is today. Oh, dear one, we're ready to leave, and we've said it before: The hardest thing we do is leave. The part that comes through my partner doesn't want to go, you see, because I'm linked to him in that way and he feels it. But the part that is on the other side of the veil that is quantum never leaves. It will walk out the door with you if you choose, holding the hand of your Higher-Self if you choose. What a gift! What an energy!

Indeed, there are those who said this would never happen, that humanity would stay static and would never move. Humanity would never evolve; they would never graduate, and they were stuck in a low consciousness. All that changed in 1987. Many of you were here, responsible for it, and you don't even remember it. Much happens on the other side of the veil that involves you. Some of you are sitting in the chairs because of it and if so, you know of what I speak.

I am Kryon. I'm in love with humanity. I am swirling in an energy of appreciation as I wash your feet that such a thing could be. I'm in appreciation that I might bring you this kind of message, finally. Congratulations, dear humanity, for you're on your way. With free choice, let it be that you would leave this place enlightened, more than when you arrived. With free choice, let it be that you will go from this place ready to do things that perhaps you were unaware you could do. With free choice, spread the light of this message and of who you are… a piece of God. I invite you to hear the music.

And so it is.

Live Channelling

The History of DNA and the Human Race

Channelled in Portland, Oregon
August 2009

Chapter Fourteen

Chapter Fourteen
The History of DNA and the Human Race
kryon in Portland, Oregon

Greetings, dear ones, I am Kryon of Magnetic Service. We would give anything to void the rules of free choice, to show ourselves in a way that would allow proof that we're actually here. But on this planet of free choice, it is you who must open that door to us. We stand like the angels we are in your lives, walking the whole duration with you. At the point in which you are born on this planet, the angels surround the area of your birth, then they stay with you until your last breath. In those beginning weeks, you might see the infant with wide eyes actually looking at the angels! The infant will point, or sometimes even smile at them, even at two or three weeks old, because the infant recognizes us. All of you did it, indeed, in those early days where there is so much change and so much to get used to being so fresh out of the womb, the angels are a comfort to the infant. Do you remember? [Kryon smile] Then slowly that reality slips from you. Slowly. But many of you have seen it when the infant looks into what you call empty space and is pleased with what he sees.

These are the same angels who are with you now. They don't age, you know? But you do. They have been with you all your lives; they walk beside you; they came in with you; they'll leave with you. And if you never speak to them, they will say nothing. For that is the agreement. But oh, dear Human Beings, if you just give us one little space of intent, and you say, *"Dear God, show me that I'm loved,"* you open the floodgate. For this is when we enter your lives to the degree you will allow us. This is when we will give you what you ask for, beginning the synchronicity, the teaching, the hand-holding, and the end of feeling all alone. This is when we fill you up to such a degree that you will not be concerned with disease or age or drama. And all you will see is the promise of who you are. This has always been the way of it since the day we began, for our job is to allow awareness that you are part of us.

The Message is Sacred

There are still those who wonder about this process of channelling. It works best when the channeller is clear. It works when the bias of being a singular Human is set aside. It is the way of it and has always been the way of it. All scripture on the planet has been written by Humans. Take a look at how this works, Human Being, for you, by yourselves, are responsible for all the prophecy and all the good things that you would find in those words that comfort you, that you say are *from God*. And so it is again that we come before you with information that is given in this sacred, ancient way.

The information given this evening for those who are in the room is given in what you would call *real time*. Those who are listening and hearing in what these sitting here call *their future* know how this works. However, this linear track of time is not our reality. We are quantum, so therefore we can see the potential of who is looking at this page, even though it hasn't happened yet for those who sit in front of me. What it means is that the message, although all about history, is still a personal one and we invite you to feel the ancient truth of it, for we know whose eyes are on the page.

All of the information I give now has at some point been given in the last 20 years within the Kryon channelling. But we have never given you this synopsis, for we are starting to summarize subjects for simplicity, for clarity, so that they begin to be less obtuse and more present in your reality. Plain speaking and plain words are what clear channelling is able to bring you.

The Information of Today's Teaching

This is the gift to us, that you would sit and let us wash your feet. And while my partner gives you information, the *third language* will be present in all the hearts here. Some of those who are listening and reading in your future will be touched as well. For we know who you are as your ears hear these words, for we see you there. We see your faces and we see the light that you have as well. It is no accident that with free choice you are willing to have this communication enter your lives this day.

Let the information given today be passed to many. This information we give now is about the lineage of DNA… from the start of DNA to the present. We will speak of the way it occurred on the planet, the way it works, the way it changed, and the things that you need to know around it in this new energy to get you to the place where you can begin to use it as it was designed.

In this teaching, we must start where my partner started even today in his lecture series. For much of what he taught in these last hours is going to be covered here. However, for those who are just joining us through transcription, we will start at the beginning… such is the way you live in a linear world. We start with history of DNA and humanity.

What's All This About DNA?

DNA is the core element of who you are, both physically and spiritually. If you had to choose a place where the Higher-Self dwells, it's in your DNA. The Akashic Record, that blueprint of everything you ever were, is in the DNA. All of the lifetimes, all of your spiritual growth, all of your talents over eons of your lives, is there. The karma that you came in with and that many of you have dismissed is there. The record of that enlightened action is there.

Chapter Fourteen

Your Human spiritual history is there, written in the quantum parts. Some of you understand this to the degree that you *know* you were Lemurian. If you feel this way, you know the incredible profundity of what's inside you. That is the reason we speak of these things. It's very hard for a Human to believe this, for it means that much of what you have felt was "outside of yourself" is actually inside. But this follows the teaching of the masters, and of Kryon, that you are a piece of the creator, and the processes of mastery are within you.

The Science of DNA

It has only been in the last few years that science has given you the proof of what we are going to speak of now. We have alluded to this in the past and now we'll just say it straight out so that there is no mystery about what science is puzzling over: Your DNA is over three billion chemicals strong. Each molecule of DNA, which is a loop, has over three billions chemicals within it. It is a molecule so small that you need an electron microscope to see it.

[As indicated before] The Human Genome Project, a scientific endeavor, has revealed a mystery: Only 3% of the DNA chemistry does anything! We speak of the protein-encoded portion of the DNA parts. It alone produces thousands of Human genes. These genes are the blueprint of life that you were looking for. But only 3% of DNA is the entire gene-producing scenario. Over 90% of the chemistry of observed DNA, therefore, is a mystery, for it seems to have no function that is obvious. There is no observable system, symmetry, or biological purpose seen within the 90% of the chemistry. It has no chemical codes as the protein-encoded parts do. Therefore, the tendency is to ignore it. It's useless.

A group of ancients makes an astounding discovery: Left on the beach by a time traveler is an amazing wireless stereo system. The small, efficient speakers are pumping out the most amazing rhythms, and the men are mesmerized by it all. They don't dare touch it, but they sit and enjoy the music, not understanding how it is being produced or anything about the gleaming equipment before them.

After many days, they begin to look deeper into the seeming magic of it all. They are curious, for curiosity is the main trait of the Human Being. They begin to analyze it. They look it over, again, not daring to touch it, and the elders finally announce to all that they have figured it out.

They see the CD player, the main amplifiers, and the processors that are sitting with the speakers... lots of equipment. The speakers, however, are where the audio is coming from, and so it is the speakers that are seemingly doing all the work.

The tribe comes together and they announce their findings: Advanced space creatures have landed and left some of their equipment. The wise men of the tribe can't figure out what everything does, but two small items are responsible for the wonderful music they are hearing. The rest of the equipment is a mystery, and must simply be space junk that isn't related to the music.

This parable speaks for itself. What is not understood as "part of the whole" is discarded due to ignorance. Hidden within this example is even more than you think. It's not just the incredible technology to play the music that is not seen. What about the creation and composition of the music? What about those who recorded it, and the incredible resources that went into even getting it to the CD player? What about the history of music itself, and the evolvement of rhythms over the centuries? There is a giant, intellectual secret hiding in the sand before them and they only see the speakers, since that is the only thing "making noise."

Evolution and what you call Mother Nature (Gaia) work together and are very efficient when it comes to Human biology and to life in general. Whether it's the appearance of photosynthesis at just the right time on the planet, or the way the Human genome has evolved, the system throws away things that are not needed. The 90% of DNA that is not understood is not junk. Hardly! Instead it is the processor and the instructions driving the part that is understood. It is processing the music that is being played by the 3%.

I will tell you what it is specifically, and in the revelation yet again I want you to ponder what it could mean to you personally. Ninety percent of your DNA is literally the quantum blueprint of your divinity. It is the blueprint of your Akash. It is the record of all lifetimes, all things accomplished, all growth, all epiphany, and all failure. For those of you who called yourselves Lemurians, it represents a vast amount of experience on the planet, all the way from the beginning, which we are going to speak of.

Things in a multidimensional state are not logical in 3D. Quantum physics makes very little sense to the linear thinker, and you are linear. Therefore, the "bias" you carry is this very fact: That your reality is based in only a few of the dimensions that the Universe enjoys. Until the "rest of the picture" reveals itself, most likely due to your scientific quests, you will only see the limited reality you sit within. Things outside of 3D will remain a mystery, seemingly to be random and chaotic instead of logical and systematic. Your DNA carries with it your spiritual blueprint, and all the instructions for who you are. All of that is in the 90%, which is multidimensional.

Therefore, the discussion of DNA is the discussion of your creation, your Akashic record, your Higher-Self, and your spiritual lineage. This is why we are concentrating on it.

Human Consciousness... 3D?

It's time to connect these things and think about the bigger picture: Within that 90% of quantum DNA is Human consciousness. The consciousness of humanity is not measurable with codes and genes. It is outside the purview of chemistry and continues to be something that science sees as a mysterious result of how biology comes together... again, there is no understanding of what creates the "whole Human." Within Human consciousness there is your ability to talk to DNA, to control it, to work with it, and to become part of it. This is what we have taught from the beginning. Therefore, one of the largest secrets of your own reality is your ability to be in charge of your own body and its basic functions.

Now, according to new science, even some of the processes of the planet itself are seen to be affected by Human thought. Are you starting to get the picture? The 90% of your DNA may actually be part of something larger than your own personal biology.

One of the first things we ever told you is that consciousness moves the earth. Consciousness is what is responsible for the vibration of the planet, and you have always been able to change what you have thought was something unrelated to you... the reality of what happens on your planet. When you begin to understand this truth, you will absolutely understand that Gaia responds to YOU!

So, these are the workings of it, spoken in a clear fashion. It's only because of the development of the Human genome and the revelation of the 90%, which seems to do nothing, that allows us to speak of what it really does. It's going to make sense to science at some point in time. The hint: When they take a look at that 90% and they begin to study it as non-coded engrams, they will begin to see that the largest part of DNA is really signal modifiers to the 3%, which is the engine that drives the biology of the Human body.

The Puzzle of the Human Body's Weakness

One month ago we spoke of this issue: Here you are at the top of the DNA evolutionary ladder on planet Earth, and you have a system that is weak... far weaker than you would expect it to be. This is a review.

Your cellular structure does not represent what was given to you originally. Rather than evolving, the quantum portion of your DNA has devolved in response to Human consciousness (the driver of your reality). The fact is that your immune system doesn't work very well! Many major diseases and viruses on the planet go right around it. Did you notice? You can't even stop a common cold.

There's a problem here, you might say, and there is! For the 90% of your DNA that's supposed to be quantum is only 30% efficient. We gave you that information some time ago, but for this lesson, I give you it again so that the logic of this teaching is succinct.

Briefly, I'll say this: How does it make you feel to know that you could have cancer in your system and your body will never tell you? [Page 178 in the discussion of DNA Layer Eight].

You've got to go to a medical expert and have a technical test to find out what your body is doing! What kind of a system is that? The self-diagnostics that have been built into the Human Being simply don't work well. Cancer is the result of an incompatibility of the Human body to cope with modern food. The result creates signals that are not congruent with how the body is supposed to work... balanced division of cells in a manner that is balanced with the chemistry of what it takes in as food. Instead, the body produces abnormal growth... tumors that eventually devour the entire organism.

Cancer is not a virus, and it isn't contagious. Instead, it's an imbalance... an allergy to modern society. The body's immune system isn't really involved at all, for cancer appears to all the body's defenses as normal cells doing what they always do. They are not, for they are specifically cancerous, but they have "learned" to hide within cellular structure. A quantum consciousness creates a system of "knowing" that alerts the body to imbalance. The result is that you would know right away that there was a problem, but you don't since the system isn't working as designed.

Another attribute of quantum DNA behaving as it should is the creation of a much, much longer Human life span. The body doesn't want to age! It wants to live! This is basic survival, so it is able, through an intuitive process, to create an intelligent scenario regarding the division of cells. Without this quantum enhancement, the body's cellular clock simply counts days and goes with the cycle of the moon. It doesn't know any better. It's not working as designed.

Some of the ancients, indeed, lived lifetimes two and three times as long as yours. It depended on where they were and how much of the quantumness they had lost. Know this: The eventual successful life extension processes being developed on this planet will all have one thing in common... the increased information to the DNA to return to a more multidimensional state. You call it "activating the DNA."

A quantum consciousness is one that is "one with everything" and would absolutely know if cells were running wild and inappropriate growth was threatening health. But your immune system won't alert you. Something is broken, but you have grown up in this reality, so the logic of what I'm saying escapes you. If you spend enough time hanging upside down, soon your body will object to walking or the whole idea of it. Eventually, everyone simply hangs upside down and starves, having forgotten how to walk. Walking is something that only the masters could do, or so the thought develops. Therefore, nobody walks. This is simply the way Human reality develops over time, and this is what we are challenging daily in this new energy.

When nerves are severed in your spinal cord, there's a chemistry that races to that area and keeps them from growing back together. Did you know that? The process itself is known by science and even has a name! It's just the opposite of what you want, and seems contrary to the logic of how a body should work. Something is broken.

Starfish can grow back an arm, and you can't. How does that feel, being at the top of the evolutionary ladder? All of this because the "blueprint" that creates the genes is not functioning as designed.

But it used to...

Let me take you back to the beginning.
The creation of your quantumness... the creator in you

Here it is, laid out so that anyone can see it and hear it and all of the attributes of it you need to hear, in plain language. One hundred thousand years ago, there were up to 17 kinds of Human Beings in development. Just like the variety of nature would call for regarding some of the other animals and mammals, there was the built-in variety that Gaia provides for in the survival of a species. Just as there are dozens of kinds of monkeys, and a tremendous variety in so many other animals on the planet, a variety of humans was also in progress. But if you noticed, there is only *one* kind of Human today. Oh, there are different colors and a variety of facial attributes, but only one kind. There is not the typical kind of differences you find in nature. You don't have the kinds with tails, and the ones without. You don't have the super small, hairy ones in one continent next to the super tall, spotted ones in another. Something happened.

Go back with me 100,000 years and you'll see that documented by science who discovered this anomaly within the last decade. They are seeing the same thing. Counterintuitive to all biological evolutionary forces, something happened to create only one kind of Human Being, and it happened about 100,000 years ago.

Now be prepared, for some of you hearing and reading this are not going to like it. This was the lecture of today by my partner, and I'm going to give it to you again for those whose eyes are on the page, or ears are listening to this transcription. Listen: A beautiful thing happened… beautiful! Something happened by design, and you were waiting for it with me, when you were creating the planet, dear ones. When you were with me watching it cool down, you knew this would happen. A divine plan was in progress.

In your own galaxy, called The Milky Way, you have a cluster of stars called the Seven Sisters. It represents seven stars, one of whom has a planetary arrangement around it. You have called this arrangement and those coming from this solar system the Pleiadians. These are the ones who visited Earth 100,000 years ago, and it didn't take them very long to get here.

This Humanoid race is semi-quantum, just like you. That is to say that within their consciousness there is a meld of 3D and quantum. There is no time; there is no space; there is no distance. They willed themselves here and they appeared. It is an advanced race that is spiritual, and it is spiritually *graduate* in nature. This isn't going to make sense to you, since it's out of the spiritual system you think is real. So let's just say that their spirituality is mature, beautiful, and totally appropriate. They arrived on schedule, by design, and on purpose.

They came to plant the seeds of quantum, divine DNA within one of the types of Human Being of the 17 types in development on Earth. They stayed for as long as it took. You should know that this took over 100,000 years, and they stayed. Slowly, all of the other kinds of Human Beings dropped away, and only one was left… the kind with the seeds of the creator. The kind that exists today.

This is the original and divine creation story given to you on purpose in a beautiful way by Spirit. It was an anointed time and I would like to tell you, souls who are here, that long before Lemuria, when you looked upon this potential scenario, you knew it was good. Intuitively, you feel it and know it, for you are firmly related to those in another star system who gave you this great gift.

The Pleiadian brothers and sisters look like you. They do not have lizard skins; they do not have strange arms and legs, funny eyes or large heads. They have no agenda and are not controlling Human thought. They're a lot taller, but they look like you! There will come a day, when it's proper and correct and appropriate, when they will show themselves. It won't be in your lifetime, for they are waiting for a specific vibration on the planet for that. But when arrive to greet you, they're going to get out of their crafts and look just like you! Then you will know that what I say here is accurate and true for they are watching as I give this message, and are smiling with the appropriateness of delivering it to those who need to hear it this way.

Listen to me: There is no conspiracy here. No one did anything to Earth or to humanity that you didn't plan. There is no control issue here and nothing is hiding. By design and on purpose Spirit allowed them, by invitation, to come and give you this gift. The only planet of free choice moved from the Pleiadians system to this solar system and literally to the earth. Biological beings slowly gained their quantum DNA and spiritually was born. Nothing happens fast with God, did you notice? God is slow and the earth is patient. But Humans have trouble with the "slow" concept. They want spiritual things to happen quickly, and the mythology shows this.

It's controversial, this information, and those listening and reading now don't have to believe it. This information is not critical to having your light shine on Earth. If it does not ring true to your soul, then pass it by. But understand that somehow there is God in you. Perhaps it isn't necessary for you to know how, but feel the creator at work in your DNA.

The controversy will rage, and it should not be forced on anyone. It is not evangelical, and it is only for you to see it as a part of the big picture. But this is where your seed biology came from and I have just given you truth.

Oddly enough, there will still be those who wish to go with the mythology that God came to the earth quickly and within a few days presented the entire spiritual system to the planet. They really don't want to think that the eye-rolling story of space aliens coming to Earth has anything to do with God, or the spiritual nature of a Human Being. It's ridiculous to them! As my partner says, *"They like their talking snake story."* Then that is what they should have, for hidden history should not be something that keeps a Human from his faith or takes away his belief.

Lemuria

Slowly, the first major civilization on the planet was born and it was called Lemuria. Know this: It was not an advanced civilization in the way you think about "advanced." But they had something you should know about. Their multidimensional DNA was at 90%, not 30% as yours is today. All the quantumness of their DNA was activated, for that is what the Pleiadians passed on to them. Lemuria was the oldest civilization on the planet, the one that was the most long lasting, which never saw war.

It was eventually broken up only because the oceans and seas rose. As I have described to you before, they became a sea-faring people and scattered to the many parts of the earth. Ironically, some made it to far away continents and science sees them as actually starting there, instead of traveling from somewhere else.

Lemurians were the original Human *society* on the planet, and they were located where the Pleiadians *originally* landed, on the top of the highest mountain on Earth, as measured from the bottom to its top... currently the largest island of Hawaii, where the Lemurian "canoes" are buried. Hawaiian elders will tell you today that this is the lineage of Hawaii, that the Pleiadians came there, for it is what they teach as how humanity began. They went many other places as well, but no society gained the grandness of Lemuria.

Lemurians had a quantum understanding of life, and they knew in their DNA all about the solar system. A multidimensional DNA, working at 90%, creates a consciousness that is *one with the universe*. One of the most ancient of your spiritual beliefs on the planet asks you to be one with everything. It's not an accident. I'll get to that in a moment.

The remnants of Lemuria are gone, covered by water a long time ago. You're not going to have proof of this, for nature has buried is completely and it is now only available to be felt in those places that were Lemuria but never covered by water. And if any of you should happen to go there with pure intent, the ancestors will come to you and will say, "Welcome home."

Here is some historical advice: Do not place so much attention on Atlantis. Atlantis was much, much later, and there were actually three of them, and much confusion around what was there, and what happened there. Which one do you want to talk about? Atlantis did not play near the role that those in metaphysics and esoteric teachings wish to assign upon it. Oh, it was important, but one of them is not ancient at all! It was so recent, off the Greek Isles, and was even reported within the history that you see today from the Greeks. Humans have a dramatic interest in civilizations that get destroyed quickly. It creates further mythology, creating ideas that Atlantis was one of the most advanced civilizations. It wasn't. Lemuria was, but in consciousness only.

Lemuria was not an advanced technical society, for it had no technical abilities at all. Yet they knew how to heal with magnetics. It was in their DNA, you see? It was intuitive information. multi-dimensional DNA produces quality intuitive information. Being *one with the Universe,* they knew all about DNA. Doesn't everyone? [Kryon smile] They even knew the shape of it... all without a microscope. That's what quantum DNA does.

The Ancients Knew!

Lemurians knew much due to the multidimensional DNA they carried, thanks to the Pleiadians. They knew all about the solar system, and about the galaxy in general. They looked at the stars and understood what was there. This created a seemingly advanced society, but without any technical advancement as you now have.

Long after the Lemurians were gone, thousands of years later, there is vast evidence that ancients still had this knowledge. Today's modern version of Human history doesn't give credibility to the 30,000 years of advanced humanity that were Lemurian. You barely even see modern Humans existing 10,000 years ago. That will change with new discovery and time.

However, what you do see is that as little as 4,000 to 5,000 years ago, the ancients still had knowledge of the stars... far more than you realize. Most of the ancient civilizations of Earth knew about the 2012 Galactic Alignment that is coming We have told you that this Galactic Alignment has already begun. You see it as 2012, and we see it as 1998. Do the math with modern computers and telescopes and you will see that 2012 is only a gross estimate from cultures without your technology. So with this in mind, you will find that you are sitting in the energy of the Galactic Alignment as we speak.

This alignment is a 26,000-year cycle that the ancients knew about it! How is this possible, you might ask? For when you rewind history to only a few hundred years ago, you find "modern science" having totally lost all this knowledge! The earth was flat, the sun went around the earth, and almost all the intuitive knowledge of how the body worked was totally gone.

The ancients had DNA that was working in a multidimensional way, as inherited from Lemuria, without telescopes or computers. They knew. They knew about the solar system, equinoxes, eclipses, and planetary movement. They even knew you were in a larger group called the galaxy. So I ask you, what happened? We reveal the obvious, that you slid backwards in a very significant way, and only recently.

When we speak this way, there are always those who want to believe that some force other than humanity was responsible for "taking your knowledge away" and leaving you wounded. This is the Human Being denying that they control the vibration of Earth. Free choice is honored by Spirit, and the whole idea of you being here is to examine where this planet will go depending on what Humans do. The splitting of Lemuria and the fragmenting of Human society created many ways to lose what was a collective consciousness. As long as the societies had a quantum community, there was a confluence of understanding. When they split up, slowly, over time, the most intuitive knowledge accepted for centuries, simply died. If you doubt that, then how do you explain that the ancients knew what only "modern astronomy" has given you in the last few years?

This is the process where the DNA that used to be at a 90% level was slowly reduced to 30%. Indeed, that's what you have today. You have literally started over. You lost the quantum intuitive knowledge, that intuition, the absolute understanding of where your seed came from, how the body works, and the Pleiadian experience was gone... lost.

Humans forgot basic astronomy, and they didn't even think the earth was a globe. All this wonderful intuition went away about the working of the Universe and the Human DNA. Only the wisdom keepers kept it. Now you have a shift at hand, and you are slowly pulling yourself up from where you are today. I'm going to tell you what is different today, and with it, we are almost complete.

The Great Shift

The lineage and history of DNA started from a precious creation that was appropriate, with the integrity of the angelic realm, and anointed with a sacred purpose. Through free choice, it declined as Humans accepted a lower consciousness state. Mythology is filled with dramatic stories of the "fall of mankind," but there were no devils involved, only the process of free choice within humanity.

You sit in a shift. You sit in an age you call New Age, but it isn't new. Instead, it is a remembrance of the ancient knowledge, and the ways of Lemuria. It is, therefore, a return to a Lemurian state, and it's about time. This is what you call the new energy, and the shift is upon you that allows for gifts to begin a slow and purposeful enhancement of consciousness on this planet. But as we have discussed, it must start within your own DNA.

Oh, Human Being, listen to me: This is the crux of why you're here. It is what we teach in this day, that the DNA, that multidimensional sacred part of you, is laying there ready to be enhanced with your pure intent. All that Kryon brought in *magnetic service* to you was to service this planet in a way where the magnetics could shift and communicate to your DNA. This new communication is a result of the 1987 Harmonic Convergence and the combination of moving into a new Gaia energy that the Galactic Alignment is creating. It gives the potentials of a return to a state that you deserve... a Lemurian state of awareness, a state where war is not an option and health is intuitive. The Mayans spoke of it! This is the highest vibration that Gaia has ever seen, and you sit in the ramp-up of it all. This was their 2012 information, and it is not taught that way by many.

Bring Human consciousness to match that vibration and your DNA will start to increase in its efficiency. It will begin to be multidimensional again. You're going to start to see it in many ways and we've spoken of this as well. Look for the signs that Humans are evolving. This evolution process is starting with what you have called the Indigo children. These children of new consciousness are very different, did you notice? It's all about how much of the quantum DNA they are using, in contrast to how linear you are! The clash between the quantum and the linear has only just begun! If you think there are issues now with the new children, just wait until they have their children! Humans are changing.

Within this "New Age," you are potentially bringing back an original energy, slowly, that was designed for today. It is time for you to return to the quantum state, which is the spiritual state that the Pleiadians gave to you, and it starts now.

At the moment, you fight a battle on the earth between old and new energy and the new energy is winning. Make no mistake that the old energy will be with you for some time. It continues to create issues that you see as problems, but are actually a "quantum clean-up." As you clean up your economy, and you're not done, you will see the ways of a new paradigm of being, of structuring, but it's going to take some time. Do not fear what it will do, for it will emerge just like you knew it would emerge... one of the strongest on Earth, filled with integrity. It will potentially represent a new paradigm that makes sense for a new age. Count on that, and Indigo adults will help you to design it. New concepts will leap over existing thought, and bring a revolution in the way life can be on the planet.

All this is part of a Human evolvement in consciousness that you can see in your DNA. *"Kryon, is a scientist going to be able to put this under a microscope and see it?"* No. Regardless of what you have been told, they will not see a change in DNA. The current microscope is 3D. So I say to the scientists this: When you develop the *quantum lens,* you will see it, for you will actually see the chemistry in the 90% random areas. It will glow under the influence of the quantumness and you will know I'm right. Then it will *change colors* with the activations you are creating and you will be able to see and track it. But at the moment, you have no quantum lens. There is no device for measuring the quantumness of the Universe that exists on the planet, but you're getting close. And when you do, the first revelation can be seen within your own biology. Here is another hint. Continue to examine things in an entangled state, for this is the beginning of the true quantum state that you can alter with 3D technology.

These are the things we wish to bring you today, and these are the things of the teaching for the day. All through this, we've been hugging you. Why do we bring these things to you? Why do we care? Because as family, you are beginning to discover a great secret… that your core souls are related to the Universe, and that your Higher-Selves are angelic and forever. Humanity is beginning to come *full circle.* Very slowly, and through the next three lifetimes, you may actually create a new earth. These are the potentials we told you about 20 years ago.

You wouldn't have missed this, dear Human Being. You are reading this because of it. You are part of the solution to peace on Earth. I know you. There will be a certain percentage of those reading and listening who, in these next years, will come see me. There will be a transition that is not new to you. It is a life transition that is appropriate and beautiful. For an old soul, it's something you've done many times, so many you can't even count. And you will come back to the planet after that, and then you will come back yet again, because you're not going to miss this! You've spent eons working on this planet, and you're not going to miss this.

This is the shift you have asked for, that you've waited for, and that you have prayed for. Many in this room will live out their normal lives, but are coming back with a higher level of DNA efficiency than you left with. We'll say it again. If you wish to see Human evolution, watch these Indigo children. For they are going to develop into an even more frustrating group than they are now! And some of YOU will be THEM, when you return.

[Kryon smile]

I wouldn't say these things unless they were accurate and true. See the clarity of the channel today. Feel in your own innate sense the truth of it all. Be aware that you are dearly loved. It is the reason we are here.

We metaphorically collect the bowls of the tears we shed in joy as we wash your feet. We invite you in the room with us now to carry on this energy of Spirit. Let us sit with you for a moment before you rise and leave. Let it be known that this day Spirit really was here… to meet the brother and the sister, to wash your feet with respect, that you would let us do such a thing as this… to be with you for an hour and to love you like we are doing.

And now you know our secret. This entire meeting was arranged so that we could just sit with you, and we have.

And so it is.

Kryon

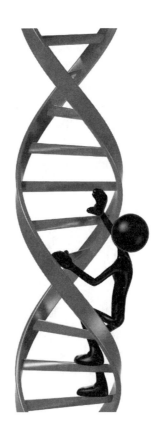

Live Channelling

The Great Scientific Bias

Channelled in Gaithersburg, Maryland
November 2009

Chapter Fifteen

Chapter Fifteen
The Great Scientific Bias
kryon in Gaithersburg, Maryland

Greetings, dear ones, I am Kryon of Magnetic Service. Quickly it occurs, does it not, the transition between Human Beings and one who is channelling a message from beyond? Perhaps it is a little too quick for those who would be in judgment about what is taking place here? But what my partner did not tell you is that there is no actual transition, since I'm always "under the surface" with him. This is his choice to have the energy of me in this way… to be able to flip back and forth in the messages of love between him speaking and me speaking. So we say to you, Human Being, that any of you can get to the place where there is little or no transition between you speaking and your Higher-Self speaking.

What I wish to speak about this day is difficult to define. I let my partner come up with a title for this channelling, for humans enjoy that. They want to realize a "compartmentalization of expectation." They wish to have some kind of identity process on everything they do, so I'll let him do it. For what I'm going to speak of is perception of dimensionality like I never have before. I wish to give you some of the mechanics of it, and a little bit about how it works, and also a little bit about what my partner calls the creation of the wild card… those things that you don't expect or don't believe in. So I have to start with an example.

Let me introduce you to Henry and Mary, they are cartoon characters, stick figures on a piece of paper. They are two dimensional. Of course, there are intelligent, since this is a parable. [Kryon smile] Their lives are not complicated, and they are simple stick figures.

Chapter Fifteen

They have everything figured out. They even have love. The two dimensionality of their lives is all they have ever known, and they are pleased with it. They know the parameters of the piece of paper they're on, and they're happy with them. That's all they have ever had. They know what they can do and what they cannot do. Henry and Mary are satisfied and content with their reality.

Along comes a free thinker, one who has been drawn a little different. This odd character begins to speak to both of them about the potential of a third dimension… the idea of "up and down." He speaks of 3D instead of 2D, as they currently enjoy. It's the beginning of the concept of a kind of reality that they have never seen, and one they don't feel they participate in, nor can they really understand it.

Let's look at what Henry and Mary do with this information. First, they don't comprehend it. It's a little too high-minded for them. Second of all, it is outside of their reality, so they're not really that interested. They don't have to use it, so to them it's conjecture and so they don't believe it really is important or exists. It becomes a fantasy of science, something that will never pertain to Henry and Mary, who after all, are 2D drawings on a piece of paper.

There are many who sit and read this 2D piece of paper in this book who are similar. Anything out of your reality doesn't interest you. Not by choice, but because you are part of the paradigm that always was… a paradigm in 3D that you have lived all your life. It's a reality that is difficult to think beyond, and many really could care less about studying it. After all, what's wrong with the reality they are in? It works.

The new shift that is upon you is one of multidimensionality. It is going to require the Human Being to understand more about what is around them, which is invisible to them, but which is very real. They must come to an understanding, and therefore a belief, that not all things are viewable and understandable within 3D thinking, and that there is so much more that is actually part of their world, but requires a logic beyond what they are used to in order to comprehend it.

The best way I can begin this study in the time allotted is to take you on a journey and give you some information. There will be some things in this message that will be interesting to those who love science. To others who are not science minded, it might not relate, but they can still participate in the analogy that I'm giving, and understand the lesson.

I wish to take you to a real place, but for today, it must be only in your creative mind. The place is real, but you can't go there at the moment... not yet. The interesting thing is that each of you has actually been where I'm taking you, when you were not on Earth. It was before you were ever Human, and we spent time here. It's an unbelievably beautiful place. The view is, shall we say, unearthly. I want to take you just outside and above your own galaxy, looking at the spiral from above. Come with me for a moment. Pretend for a moment that the pressures of space and the temperatures don't apply to your Human body. None of those things matter, for you're in a protected bubble that is your spiritual self. All together, we go and we will watch this magnificent sight.

As a Human Being, you're struck with the silence of space, not understanding or even appreciating the fact that every single star sings a song. I hear them all. Silence to you is a symphony to me. For the vibratory rates of the light that is emitted from the stars all combine into a chorus, a manipulation of vibrational sonority that is beautiful. The universe sings to me, for I am quantum. The parts of you that are quantum are beginning to broach past the three-dimensional parts. That meld, that confluence, is going to create paradigms of thought that are different from any others on the planet. For there has been no time on the planet like this one, where you're asked to think out of the box of your comfortable reality, and go beyond the wall of your natural bias. Look at your galaxy with me for a moment. The beautiful spiral of it, is all moving slowly together as one… rotating slowly like a plate of lights. Take it all in.

I give you science today. I give you knowledge today that will only come about and be known within your future. And because of the transcriptions that are taken today, there will come a time when you will point to this particular message and say, *"Kryon was right."* And when you do, when the science confirms what I tell you today, I want you to look at the entire message. Because then your belief factor will also know that I speak truth when I speak about your relationship to Spirit, to the creator in you. I'm right when I talk about what is in front of you and your future, and the only reason I give you the science now is because your linearity and your bias will connect them in the future. *"As goes one, goes the other,"* you'll say, *"therefore I will pay attention to all that is said."*

Looking at The Galaxy

It's beautiful, isn't it, as we are suspended here above your home Galaxy? In what you call the silence of this moment, looking at the galaxy moving so slowly, spiraling all together as one, it's unbelievable, unearthly, spectacular beyond words. Now I take you inside and tell you a little bit of what's happening that is a mystery to your science. There are odd things out here that don't fit earthly paradigms or rules. They don't fit *your* physics.

Henry and Mary, the two stick figures on the page, had scientists also. They had their 2D laws for physics, and that's all they needed. Everything worked just fine as long as they stayed on the page. You have four laws of physics, because you're in technically in 4D, and those laws work just fine. Those laws have proven themselves over and over, and as long as you stay in 4D (which you call 3D), they will always work.

Here's an esoteric question for you. Take a look at the stick figures Henry and Mary. How many laws of physics are there really for the 2D characters? Is there a whole set that encompasses multidimensional reality, or just enough to satisfy 2D? The answer should be obvious. Physics is complete no matter how you perceive it. Therefore Henry and Mary are only aware of, believe in, and are using, 2D, but *all* the laws are still there... ready to be discovered. Colors are still colorful while being seen by those without color receptors. You had black and white TV for years... yet, everyone "knew" there was color in those pictures, didn't they?

So extend that thinking and let me ask you this, three-dimensional creature: If I told you there were six laws of physics, covering a dimensionality that you don't see, how many are there for you? The answer is the same as for Henry and Mary. There are more than you know about, even if you are only aware of, and actively using, the few you have.

Do you see? The four laws you have, work fine. There is nothing wrong with them, but there are more, and that's why we take you here above your galaxy to show it all to you, and to present something that astronomers also can see. Look: Something is weird with the way the galaxy is moving. Did you notice?

We've given you the two additional laws of physics before, and this is not the time to explain them again. But when you get into an interdimensional realm, you're looking at multidimensional energies that must contain more information than your current physics explains. You have four laws now. Call them Newtonian, Euclidian, Einsteinium if you wish. These are the ones that brought you to where you are today.

But now gaze with me at your spiral galaxy for a moment and watch it move slowly. It does not move like your solar system.

The laws you have of *objects in motion* carry with them a three-dimensional bias of consistency. Your science looks for empirical laws and they find what they believe is true for everything. But what they don't realize is that there is a bias applied. It only works in one direction... in 3D. If you apply the rules only on that one playing field of time, you can apply linear mathematics and figure out what you need from that. It's all in a straight line, all forward, never changing, always the same. You might say your science is *biased in simple consistency!*

"Kryon, what's wrong with that? Sounds find to me!" Here comes the free thinker who is saying: "Interdimensional things do not apply to 3D logic or bias. The laws of interdimensional weak and strong forces are beyond 3D understanding, and may even seem to be chaotic and inconsistent."

Let me give you a further explanation. Your solar system works like you expect it does. Within the kinds of physics you have applied to the way things move in space, you have objects that are closer to your sun that move faster, like Mercury, for example. Then there are objects further away from the sun that move slower (the outer planets). The laws of orbital mechanics are in play. The distance from the sun develops into the 3D laws of orbital mechanics based upon the rules you have discovered for gravity, mass, distance and speed. And the rules are correct... for 3-D. Again, it lets you send spacecraft to the planets, to be so precise, to meet them in orbit, to take pictures and analyze them.

But look for a moment with me right now... *this is not the way your galaxy is moving.* It's in an elegant motion that defies the *law of the inverse square* (a law that defines how energy dissipates with distance from the source). It defies the basic laws of gravity and force. It defies the simple, biased, singular attributes of the way things move in space. Look at your galaxy with me. Watch it spin. It's almost like it was on a platter. Everything moves together. Everything! It's all rotating at the same speed, relative to the center… like a giant wheel that is all connected. This giant platter behaves like all the stars are pebbles, and are somehow glued to the fabric of space, all moving together. Now, the spirals and their sweeps will tell you that it wasn't always this way, but why is it now?

How can that be? What are you seeing? Let me give you a hint and a clue: We've spoken about the weak and strong interdimensional forces which are the undiscovered laws five and six of physics. The way your galaxy moves is all about what's at the center of the galaxy, and displays these forces. You think it's a black hole, but it isn't. There's far more to it than you would imagine. Have you noticed in physics there is always polarity? From the smallest atomic structure to the largest, there is always polarity. You also see this in magnetics. It's also hiding in gravity. It's a staple of energy everywhere, everywhere. There are always two kinds of energy, and they work against and with each other to create dimensional reality. Matter itself is one polarity of reality, and anti-matter is the other. Always look for the push and pull, for it will show the way to the answers to the most perplexing issues of physics.

At the center of every galaxy there are "the twins." The twins are in the middle of the Milky Way as well. You've got two energies: One pushes and one pulls. However, you see it in your perception as one giant Black Hole, and you have an impossible name for it: "a singularity." You assume the gravity of the Black Hole is somehow gripping that spiral and making it spin together in an unusual fashion which violates all the laws of Newton. It's not so. What's happening in the center of your galaxy is beautiful. It is an elegant interdimensional force that is not gravity, which spreads through the entire region of your entire galaxy, a force that glues it together in a way that you do not have laws to explain... yet. In addition, there is something hiding that science is only now beginning to wrestle with.

All this explanation, to get to a place of logic for you that will broach a very big issue. Simply stated it is this: When you step into interdimensional physics, and this includes the energies of what you call spirituality, you will find something you didn't expect: Consciousness... *physics with an attitude*. The interdimensionality of your galactic center has consciousness. It has to. Anything interdimensional is aligned with creation. I'm speaking of things you don't understand. These are high minded, sometimes unbelievable attributes that haven't really been broached in this way before. When you break the linear logic wall down from what you expect in linear physics, you're going to run up against things that don't make sense to your bias. They won't make sense... not just because they're in a quantum state, but they contain something else that "consistent 3D science" does not wish to accept... intelligence in physics.

Your science is very proud of the Big Bang Theory. They have it all figured out and they have a timeline for it. This is really funny to us! How can you have a timeline for a quantum event? There is no time in a quantum state, yet they have it all figured out. They've even figured out that there's a *residue* they can measure that proves they're right. How clever of them!

Let me ask you something, if you smell that wonderful residue aroma of bread cooking in the kitchen, what does that tell you? Does it say, "bread was cooked here four billion years ago" or does it say, "it's being cooked now"?

It's the bias of straight line thinking in a singular time dimension that smells the bread and calculates how long ago it was cooked! There is no understanding that the quantum event of the "Big Bang" is still happening. It explains the energy of Universal expansion. It even begins to explain the "energy of what you can't see." The "residue" they measure is the proof of the reality of an event still in progress as you see it in 3D, but an event that is the *reality of creation,* within a multidimensional state. It has to do with the *crossing of dimensional membranes.*

Look at what a 3D mismatch the current theory is: How could everything have come from nothing, then at a speed greater than the speed of light, instantly expanded, violating every law of current physics, to create the current mass of the universe in a nano-moment? Yet the bias of singular linear thinking lets all that happen in the time line of an instant... and they have it all figured out. They should all be celebrating with Henry and Mary! [Kryon humor]

Let me tell you something I have never, ever described to you before. The center of your galaxy spit out the matter that is you. Science has it backwards. The twins in the center of your galaxy lead to the twins in the center of all the other galaxies. Millions of them, billions of them. They're all connected in a way you cannot fathom outside of space, outside of time, like strings between friends who have consciousness. Not the kind of intelligence and consciousness that you see in your brain, no. Instead it's a benevolence, an intelligent glue that postures the universe in love. I told you that you wouldn't understand all of this. This is high-minded, high-thinking, and many are simply not ready for it.

The Gaia Effect

Let's go to something else. Life on the planet, and the way it was created, has become controversial because there are those in science who must linearize it all. Darwin gave you the possibilities of an evolved life system. He showed how it might work, perhaps, in a random selection of biology over and over through billions of years, creating what you have now. But then, enter *The Gaia Effect*.

Scientists are looking at earth history and they're starting to see something very bothersome to other scientists: There may be a consciousness that has created life. Naturally, true science does not want to think this way, for 3D straight line thinking of your current science does not allow for rules outside the box of total consistency. The real irony here is that *singular consistency bias* does not allow for *creator bias*. Could the universe be biased toward life? In this irony, the Human is biased due to limited dimensional thinking, and the Universe is biased in love.

The controversy goes like this: Earth history shows that life continued to be created and destroyed on the planet through four billion years. It started and it stopped and it created itself and destroyed itself over and over. Whereas life was once looked upon as an "against all odds" attribute of the planet, and nowhere else in the Universe, it now is seen as having been created again and again!

Some say, *"Well, that's a random event happening."* Really? What are the odds after life had destroyed itself, of the that incredible randomness striking again? How's that for evolution? Something that didn't work… returning! What do you think about that? Scientists are beginning to consider *The Gaia Effect*, as a consciousness coming from somewhere, somehow, that is biased to create life. It's outside of the purview of what you would call chance. Over and over it happened, until the planet it got it right. Photosynthesis was the answer, for it created a balance… plants and trees to consume the by-product of life. So finally, the balance began.

It took a long time for that, but life was always created again until the "system" got it right. Even when the system snuffed life out, it returned! Even when the earth was barren of life because it hadn't worked out, it was created again up to five times. Science is starting to see this and is wondering how it is that earth seemed *biased to create life*. Some say there's a consciousness; some say that is not true… there couldn't be, just couldn't be.

But there is, dear one, and it's an interdimensional consciousness that glues things together. Because when you get into an interdimensional state, you're starting to touch the face of God, the creative energy of the universe, and one that is indeed, biased in love.

Geological Surprises—Rethinking time

There are those who study the way the earth geologically came about, and again, in their straight-line thinking, they are biased. They are biased because they look at erosion patterns, they look at the way things used to be, then they apply specific universal laws to everything on the planet from then on. Well, there are some surprises: Have you heard the latest? How long did it take to cut the Grand Canyon? How many millions of years would the water have to trickle by in order to cut the canyon as it stands today? One million years, two million? It's still posted on the placards on the historic sites of how long it took. But now geologist are starting to change their minds because they have discovered other attributes that don't make sense. Now they're assigning a timeline of approximately three hundred years!

What happened to their logic? What they're now seeing is a *wild card*. That is to say there was no trickle of water. Instead, there was a sea that emptied into it… a ferocious torrent of water that cut the rock over a much shorter period of time… and it flowed in both directions due to massive land upheavals… mostly due to the weight of the water. Add to this was the fact that the west coast of the United States was next to it once…. then it wasn't! Does that tell you there might have been some upheavals? They also assume that the land was flat during all this. What an assumption! What if it were downhill to the degree of thousands of meters difference between one end of the canyon and the other. Do you think that might create stronger pressures? Think "Niagara Falls!" Look how quickly Niagara has cut a path even in your own lifetimes! Water under pressure will do that.

But geologists think with this bias: It was always serene, a river, and no real evidence of cataclysm. Really? They should have been there to see a magnificent sight!

Outside the paradigm of thinking, it is, and accurate it is. You see where I'm going with this? It is the consistent biases that keep you in a straight line rut like the 2-D figures on the piece of paper. You've got to start thinking out of the box and look for what else there might have been. So I've given you the Grand Canyon story, so that you can absorb the next one.

The Unspoken Geology of Lemurian Existence

I've told you some odd things about the planet, and geologists always roll their eyes. For Kryon gives you information that is often "geologically impossible," they say. I told you about Lemuria. I've told you that the original Lemurian civilization was centered on dry land at the base of the highest mountains on earth, measured from the bottom to the top, which is the Hawaiian mountain. It is one big mountain with several peaks, and the peaks are what now stick out of the water, which you call the Hawaiian Islands in the middle of the Pacific Ocean.

In the days of Lemuria, we told you that the land around the base was dry. Geologists laugh. It's in the middle of the Pacific Ocean! How could that be? I'm going to tell you how it could be. First, you must understand that geologically, 50,000 years ago is not a significant enough time for something like plate tectonics to have much of an impact on this attribute. 50,000 years is actually very little time in geology… yet the water level at that time was more than 400 feet lower than it is now. This is because you were in the process of a water cycle which we have discussed before. So that's one of the attributes that came into play. However, the other attribute is the biggest reason, and one we have never spoken of before.

The mountains of Hawaii slowly move over what is called a *hot spot*, that is to say it slowly moves over a tremendous volcanic core of activity that has existed there for millions of years. 50,000 years ago this *hot spot* was in the process of a giant "bulge" that actually gradually lifted the floor of the ocean around the Hawaiian mountain, more than six thousand feet. That is to say the mantel of the earth bulged enough due to the volcanic and tectonic pressure, to lift those mountains higher than they are now, to create a large area of dry land that contained the mountain of Hawaii.

When the lava pressure was released into the mantle, and it was, the bubble slowly deflated. This took several thousand years, without a cataclysmic eruption, as it was slowly released itself both on the peaks of the mountain and poured into the sea, and into the mantle of the crack that is today one of the tectonic scars of plate motion. This created a situation where the base of the Hawaiian mountain was above sea level for a while. The slow release caused the bubble to subside, and Lemuria slowly was flooded. This is the story we told you originally, and the reason the Lemurians became sea faring, and moved to many other places.

Convenient, it is, that all the evidence of Lemuria has been destroyed as it should be. It makes you wonder. It makes you wonder if these things might be so. But a linear thinker will tell you that this can't be so, since they have never seen evidence of it in history. You haven't seen that kind of bulge before, therefore we have the same "Grand Canyon" effect, where the truth lays hidden due to biased consistency. What you never saw is therefore not possible.

By the way, the evidence is indeed there of the bulge, for the striations of the ocean floor still show an odd symmetry around the mountain, giving hints that it was once stressed upward through volcanic influences, and then subsided. There are also bones of animals at those depths, buried deep, that would tell a biologist that what is on the ocean floor in the middle of the Pacific Ocean, was once exposed to sunlight... about 50,000 years ago.

Creation—The Proof is in the Odds

Let me tell you about creation. Astronomers are starting to talk about *Intelligent Design*. Now we're getting somewhere, since they're starting to understand that the quantumness of the universe might indeed have consciousness. Against all odds, you live in a parameter, an attribute in space that statistically continues to be "against all odds." You are in a universe created for life! If you could throw the dice of physics and create a universe, it would never come up this way. Never.

Statisticians have said that it's *out of the possibility of chance...* yet you sit in an earth teeming with life. You sit in a universe teeming with life. There is life on the planets around you, but you just haven't found it yet. Microbial it is, and it represents the beginning attributes of single-celled life. It's all there. You'll see. Take a trip to Europa, and look around a little in the ocean. You'll see. Life is absolutely the way of the universe... everywhere. You'll see. And it's against all odds that it happened, and science is now seeing that. It is so out of the statistical model of the creation of any universe, that they have labeled it *Intelligent Design*. There has to have been a plan.

In the middle of your galaxy, the twins exist, pushing and pulling interdimensional energies that literally have an intelligent complement to them. All of the stars move together in unison with it. Forces that are way beyond gravity are involved. It's an interdimensionality that glues the galaxy together, and that is something I want to speak of because it has to do with your future.

Your Future

The futurists of your society have a tendency to look at *what was*, and then project *what will be*. Do you see the straight line thinking? In a bias of consistency, they say, *"Because of this, therefore, there will be this."* They're looking at a consistent model of old energy, and in their projections, never giving it a chance to change. They are denying the very ability of Humanity to move past *what was*.

What is it that has been consistent on the planet regarding consciousness? Let's name the attributes: War, poverty, suffering, drama, a repeat of the same, a repeat of the same, a repeat of the same. Fractals of time that come and go and come and go, which create a consciousness that repeats and repeats and repeats.

Now you can see the bias, and why it's there. For anything that stays in the same cycle is expected to stay in it forever. This is also echoed in your physics. The more multidimensional you become, the less consistent you will be in thinking this way. That is to say, eventually you will *expect* things to happen that have never happened before. You are in a shift that is allowing for it, and all around you, things are changing. Listen: You cannot apply *last year's rules* to peace on earth. Is this possible... "expectations of a wild card?" Yes... more than possible, it's the most probable outcome that we can give you.

What do last year's rules tell you about healing in your body? They say it's incurable? It isn't. What do last year's rules tell you about the fear that is being generated over the shift you are in? Does it tell you that you're going to be covered by water. Well, if you apply the rate at which things are melting and then you project all this into the future, you apply the discourse of poor science... then of course you will be! But that's awfully consistent of you, isn't it? You see what I'm saying? You're not giving any allowance for the *wild card*. You're not giving an allowance for the fact that this planet is moving into an entirely different magnetic attribute. The sun is cooperating; the universe is cooperating; it's almost like the twins can see who you are!

The twins are physics, a magnificent pushing and pulling of an interdimensional attribute which we cannot explain to you. The twins are not God. The twins are a result of a God-biased universe, and an interdimensional physics that you are only just beginning to discover and question. Your science is only beginning to see it. The Gaia effect, Intelligent Design, the way the earth was put together, the way there were no accidents that you're here, against all odds… that ought to add up to something in your logic.

You are in the middle of a tremendous shift that you've asked for, that it's time for, that I came for! It's not one that's going to destroy you! You don't have to fear it. It's one you're in control of as you're becoming more multidimensional, and therefore more quantum. You're starting to understand that the consistency of life as you have seen it is truly is an old paradigm. The consistency of fear, of hate, of disappointment… is beginning to change. If you want to be consistent with things, know this, that the love of God is the most consistent and persistent thing on this earth and in the universe. Consistent, it is! So persistent that it would not stop until life was created, and until love was discovered.

The Consciousness of Kryon

We wanted to paint a picture today. It was mostly scientific, partly a logic puzzle, to show you that your current 3D logic is not expanded. It's not the kind of logic you're going to need to move into the future. We told you to *expect the wild card*. Expect things that have not happened before. Change history by thinking about future things which may be evolutionary and represent wild cards themselves. Can you recognize some of those wild cards in your very immediate past? How many times do we have to give you these things to look at? None of the quatrains of Nostradamus are accurate today. Books have been written about the tremendous upcoming war with Islam. Well, someday you can read them and laugh, because it's not happening that way. That entire scenario is based upon a consistent, unchanging consciousness on earth, where everything repeats in an expected old pattern, and it isn't!

Are you the expected pattern on earth, or are *you* the wild card? You see, there are things coming that you didn't expect. I know these things for I see the potentials that are not openly available to you. It's the consciousness which is "baking in the oven" that is working right now in ways that you do not expect. This is not future fortune telling. It is instead just stating the facts of the potentials of the minds on the planet... where they're going and what they're thinking.

I want you to leave this place with hope. We saved this message for this particular group. It's an advanced message, and it does not fall on the ears of those who were just here for the first time or in the energy for the first time. It falls on the ears of old souls. Don't you remember this? Remember this with me, for you expected it. That is why you are here, and that is why you will return. Now you know: you have something in common with those who have made it this far in this book!

Finally, this: I was with you at the wind of birth, each one of you. Before you slipped into this planet, before the angels stood around the bed and sang to your mother at the joy of your birth, I was there. Right before you came again, I inquired like I always do, *"Is this really something you want to do? Look at the potentials, and the hardships, the disease and sorrow of being Human. Do you really want to go back?"* You looked at me like you always do and said, *"Send me in. I can hardly wait to get back and finish what I started."*

Now you have some of the first signs you've ever had since you were a Human, that this planet can move into graduation, vibrate higher and become part of a confluence of energy that you've only dreamed about… and you're sitting on the edge of it.

I am Kryon, lover of humanity. This is the truth I give you today. I give this message in order to instill in within your hearts and in your minds, the hope that these things are so and that they are true. Perhaps some of you today will leave here differently than you came because of this.

I am Kryon. And so it is.

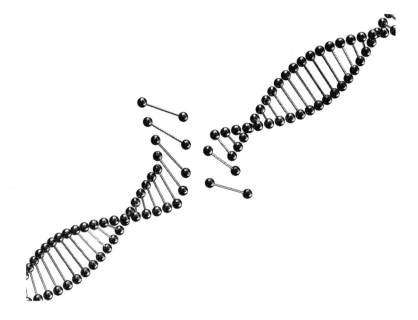

Chapter Sixteen

Current Events

Kryon
Book 12

Lee Carroll

Chapter Sixteen
Current Events
by Lee Carroll

The Great Intellectual Transplant Debate

Okay, it's time to offend people again. Don't try to figure out the workings of Spirit and place them into a Human box! Why do we place the creator into our scenario of laws, expectations, beliefs and even our 3D logic? I wish I knew the answer, but we do it all the time. Therefore, there are always questions that ask, *"What happens when (this or that)...?"* These intellectual logic puzzles become challenges to our spiritual beliefs as though God were in a "rule box" like men are, unable to escape a seeming conundrum of logic. Perhaps these things need to be discussed, but I wish they didn't. It shows the foolish Human 3D bias very strongly.

It was 1967 when Dr. Christian Barnard, a renowned South African surgeon, did the first heart transplant of all time. There was a flurry of congratulations and a huge outcry from the spiritual community of "foul play"! There were death threats to Dr. Barnard for interfering with God's law and of mixing the soul of one with another. It was seen as wrong, and again science was seen as playing around with life—a spiritual no-no.

Today we laugh at the naivety of all this, for this life-saving procedure is performed as many as 40 times a week in the U.S. alone! It has become something that is "normal" for Human culture, and most societies have it perfected. The religious objections have all but disappeared, except from the very fundamental types who won't allow medicine even for a dying child. So what happened? I would like to think that consciousness started to understand the meld that our medical procedures have with an evolving humanity.

Doctors and life-saving technical processes are not against God, yet many still put them in that category. I firmly believe that what we are given to extend our lives is given right from the creator and is given in love. If a procedure saves the life of a loved one, can you thank God for allowing that doctor or that process to be part of your evolved generation? I can, and often do.

I believe scientific discovery is appropriate. Again, it's what we do with invention and discovery that is the free choice that we have and is the measure of our enlightenment. Almost every core physics and medical discovery contains profound choices: death or a cure, a power source or a weapon? It's the way of it. I thank God for what we are discovering, both in the scientific world and in the spiritual world, for these things all belong to the creator and we are simply discovering what was always there, slowly.

THE QUESTION: *If each piece of DNA contains our personal spirituality, what happens to the Human Being when a transplant is given of a Human organ? Suddenly, there is the core Higher-Self of one person, inside another! How can you mix Akashic Records? What does the other DNA think of all this?*

First of all, let's look at what happens medically. The organ is rejected! That should tell us something, I think. Indeed, the biology of the body recognizes the cellular DNA is from another Human Being and does not belong in the system. There really is no other compelling medical or biological reason for the body to reject an organ with the same blood type and that has seemingly compatible biological attributes, but it does. That means it "knows."

So, is that a sign that transplants are wrong? Kryon has actually answered this and begs us not to intellectualize everything into men's logical thinking, for there is "spiritual allowance" for technical invention, actually given to us by God. It's an age-old argument by spiritualists, who have decided that we should stay as we were born, and that if God had meant for there to be pain killers, we should have been built with them inside us.

Kryon answers this, too! He says that as masters, we indeed have it all within us, and sometimes invention is actually part of the plan to "wake up" a system and start it evolving.

Basically, God is not in a vacuum, or in the dark about what we are doing, uncaring about what is happening, or filled with rules about what we can and can't do that will somehow violate the whole spiritual experience. Instead, Kryon tells us that many times the availability of a donor organ is manifestation at its best, and that the whole process is sacred. God knows what is available to save your life. Sometimes you can heal yourself and sometimes you just can't grow a new kidney or liver in time. So there is understanding and appropriateness here, and God is right there with us to help extend our lives with a transplant.

On a more esoteric note, Kryon has even indicated that you have the ability as a receiver of an organ to slowly change the donor DNA over time into your own! As the cells divide slowly, they reproduce and regenerate the blueprint of your own DNA! Is this possible? Kryon says yes, but it's advanced and requires a consciousness that can "talk to the cells" as we have discussed in many seminars, and which Kryon speaks of regarding DNA Layer Nine.

It also requires the ability to "mine the Akash," as Kryon calls it, a process that is a new tool of the New Age and that sees all our lives as melded and available, including the current one. He invites us to do this, and as proof of the fact it is happening, to slowly be able to reduce the drugs given to prevent the rejection that transplant recipients must take—all this with the doctor's permission and encouragement, or course.

But what is really happening when you put the organ of one person, dead or alive, into another? Is there any kind of a meld? What does the DNA do? What Kryon tells us is that the DNA "knows" what is going on. Remember that? This is the hardest thing to relate in an intellectual discussion, for we think of the Human chemical system as reactive and static. We don't consider the spiritual side or what this book is teaching. We often don't consider DNA as "knowing" or having intelligence, but it does. As Kryon has indicated, the DNA field is what creates Human consciousness, not the brain. Therefore, it "thinks" and it is also our "knowing." It knows what we know and more.

This means that the DNA in your body "knows" what you are doing and knows about the transplant. It knows to be ready for the jolt of the incoming organ. It also celebrates the fact that it may extend your life. The Human body's survival instinct is the strongest one we have and, believe me, the DNA knows what you are trying to do.

Initially, the 3D chemistry will reject it. This is simply because it is foreign DNA and in the old energy of humanism, there was no way transplants were possible. Kryon even tells us that as our civilization continues to evolve, science will see "diminishing rejection." In other words, for no apparent reason, Human cellular structure will not reject the new organs to the same degree, indicating a "learning curve" of biology in general about modern methods. Sound too weird? It's spiritual evolution, DNA creating new genes that will enhance a modern process, and we are going to see it.

If DNA is aware, and it is, and if the transplant is going to save your life, which it often does, then the answer is clear: It is appropriate and of God. In addition, the body "knows" about it and works around it. The DNA field changes to compromise with the biology and the new organ, even though it is being rejected in a 3D chemical way, becomes integrated into the Human's Merkabah (DNA field).

Sometimes this, indeed, creates interesting phenomena, for the Akashic Record of one is placed into the other and often there can be a combination of apparent consciousness. Sometimes it's subtle and sometimes not! Of all the things in measured science that I can point to that would prove anything this book is saying, I point to this: Why would an organ transplant recipient suddenly take on some of the attributes that the donor Human had? They often do!

Documented in many cases, the host sometimes starts to like the favorite food of the donor or has some of the habits of the donor. Now, where did that come from if DNA is simply a chemical producer of genes? It has to come from the Akashic Records, which occupy 90 percent of the DNA molecule. In the process, this also alters the very consciousness of the Human Being who has received a transplant from another.

Ah, but the intellectual, esoteric arguments continue. *"Well then, what happens in the Cave of Creation? What about the soul? Whose soul is seen then at death? Do the two souls in the host's body now reincarnate together? Will this upset the spiritual balance of humanity somehow?"*

To Spirit, this is all so funny, since there are profound and beautiful answers to all these questions, yet they are asked as though God is somehow at lunch and the Humans are running amok and now the system is out of whack. This is the incredible Human 3D bias, which can only judge things by what is knows and not by what it cannot conceive of.

Chapter Sixteen

SUMMARY: The transplant process is *known by God*, and also by your body's DNA. It is not spiritually inappropriate. The core Higher-Self knows about it, too, and begs you to integrate it over time, eventually without all the drugs. It is celebrated by the rest of your DNA as something that has saved your life. The doner's DNA also celebrates, since it "knows" too! Life is sacred, and together, you and another have come together to make is all work.

Survival is sacred. God isn't shocked, all is well, and the angels celebrate what you have done, every day of your life.

Then, or course there are those who say that because of it, you are now damned forever, and that Heaven can never take you. Even Hell won't recognize the DNA since you messed up to badly. You will have an absolute miserable life, and some churches won't let you in the door either. A transfusion often carries the same damnation. Sorry... just had to let you know what Humans do with things.

I have a favorite saying by a famous musician from Texas named Butch Hancock. It kind of says it all:

"Life in Lubbock, Texas taught me two things:

One is that God loves me so much that I'm going to burn in hell.

The other is that sex is the most awful, filthy thing on earth and that you should save it for someone you love."

Thanks Butch!

Artificial Life - DNA of a Different Kind

In the year of the publication of this book, Craig Venter, the man who mapped the Human DNA genome in record time, created what appeared to be a new life form. Controversial, of course, but scientifically profound, his institute has created a very simple DNA that reproduces itself and seems to "have life" by the standards we have used in science for decades to define it.

The Human Genome has more than three billion parts. His new life form is quite simple, with only a million or so parts, and all of them are protein encoded and gene producing. His DNA is core life and may actually someday be seen as the Human's best life-friend. It is designed to eventually help carry intelligent healing properties from one Human DNA molecule to another and could be considered the first "intelligent nano-robot" to assist Human life. The possibilities are endless for virus fighting and nano-healing.

Naturally, there is danger here, and any scientific discovery this profound has a good and bad side. Who wins the controversy and the appropriateness of it is not the subject of this book. But still there will be those who want to know what Kryon sees here and what it all means. So I will let you know, since Kryon spoke about it very quickly after it took place.

First, don't fear it any more than you fear genetic alterations, which have been with us for decades and which you participate in willingly or not. We must be careful and respect it, for it also might be the vehicle for terrorism. So it's back to the Human intent issue and what we do with things we invent. Nuclear energy can help clean up the air or destroy humanity, our choice. The most profound inventions of the centuries can be used in either direction. So we are again in charge of our own destiny.

But don't do the "evil" thing and attach forces of evil to it or to the new life form. It's simple, not very smart and very programmable. That's why there is only a fraction of the chemistry of the Human Being. It's not evil, any more than was the first heart transplant we spoke about earlier (which was considered the work of the devil). Get over all the drama around the "devil within it" and see it for what it is—a chemical invention without consciousness. Then you can decide the ethics of it for yourself, and agree or disagree about its right to exist.

This new DNA does not have a quantum attribute. No Human Being can build that, since there is no one on Earth at this moment with anything that can even measure an interdimensional field, much less make a complex one out of 3D chemistry. The artificial life is, therefore, only chemistry and does not have and never will have the same kind of quantum chemistry that a sacred Human has.

SUMMARY: Don't fear this invention. Kryon speaks of our kind of DNA as the only sacred DNA on the planet. Original DNA is a product of natural chemical evolution on the planet; it happened naturally. Venter has duplicated the chemistry of that evolutionary process, but it's the Pleiadian story that gives us the sacredness of "spiritual creation" and is the "spark of life" that is our soul consciousness. It's not basic chemistry that gives us the love of God within. It's the sacredness of the "creator inside" that is the key. It's quantum information, not chemistry.

The Last word
The Great Human Commercial Machine
... even in the New Age

There are many who are claiming that there are now 14 layers, perhaps even more. They give even greater esoteric meanings to them and names that are their own. Bless them! For this is simply an expansion of what this book is teaching! Remember, DNA does not actually have layers! It has groupings of information that can help you to understand a multidimensional engine that is interactive and really has no linear parts at all.

So when you hear someone teaching more layers, pay attention. Discern for yourself, for you may be looking at the next evolution of how DNA is progressing on this planet. Get used to the fact that there is no "he said, she said" in all of this, where one teacher has more layers, different names, and is therefore the "correct one." If anyone says that, they have totally missed the point, for this is a study of the creator within, and that creator does not have parts you can count. Instead, *celebrate new information*. See if it instead fits into existing information, but is just stated differently. Perhaps it's entirely new? In any case, your DNA is "aware" *and able to give you a firm "yes or no"* on just about any subject, just like the master creator's mind can. It's time for New Age Humans to weigh these things *out of linearity*. Look at them energetically and feel the truth or don't. It is *time to discern* instead of simply believing everything you read (including this book).

In 1999, Jan Tober and I published a book on the Children of New Consciousness, called *The Indigo Children*. It's a report on the evolution of the Human spirit. The purpose was to help kids and parents all over the globe deal with seemingly difficult kids, and to explain what was happening. It was a hit, and over half a million of these books have been distributed in over twenty languages worldwide. Parents and teachers related! We subsequently have done two more Indigo books, filled with more information for parents, educators, and managers in the workplace on how to deal with these exceptional conceptual children.

But back in 1999, as soon as the popularity of the subject was seen, there were immediately books about other "new" children with other names. It was funny. Almost as the ink dried, there were new books claiming other kinds of kids with different names! [not Indigo] It was almost as though everyone wanted their own slant with kids with names they had "channelled."

The ironic part of this is that the Indigo name was not originally channelled information at all, but a result of the visionary Nancy Tappe who had medically validated *synethesia*. One of the attributes of this medical anomaly is the ability to sense colors around energy. She *saw* the indigo color around newborn children, so all new children were called Indigos. This means: *"all children of new consciousness."* It's not about an esoteric color seen with psychic abilities. So by definition, all the children being born were in this category of new consciousness. Nevertheless, it seemed like there was a "war of new kids' names and colors" as more and more jumped into the fray, claiming that theirs were different, not even bothering to check out what the name really meant.

A movie was made called *Indigo* that had nothing to do with Indigos and didn't help one kid on the planet. A *documentary* was made on Indigo Children that didn't even mention Nancy (or her contribution to seeing these kids and what it meant). It was a free-for-all of those who could capitalize on the Indigo subject, and get their name up front too, and they did. The disappointment is that within all this, I feel the children were left behind, for these projects really had very little to do with kids. They instead took advantage of the popularity of the name and missed the entire purpose of the original information.

Now we have the potential for it to happen again, for this book will spark many others who claim the DNA names are wrong, or that there are more or less than 12 layers, or that the whole thing is part of a giant conspiracy to capture your soul. Those who know Hebrew will argue with the meanings (they already have). They will wish to add to this book, replace it with their own information, correct it, or perhaps even to burn it. I don't care. What I do care about is this: Will the new publications to come further your own personal study of the creator inside? Will it create peace where peace was not? Will it stir up drama or will it soothe the soul?

I wrote this book to expose the greatest evolutionary shift in Human history. From 1987, as the galaxy has crept at a dead snail's pace toward the galactic alignment of 2012 *(.001 degree per month as we move toward the cross hairs of a mathematically disputed line-up of our sun with the galactic equator and the center of the Milky Way)*, we have been seeing tremendous change. We are discovering that we are not victims of the earth. We are questioning the very premise that kept us in the dark, that we were born dirty and didn't have a chance. Instead, hundreds of thousands are awakening to a potential that they have the creator inside them, that they can heal themselves emotionally, physically, and start an informational flow to their own bodies and the planet that can eventually change how people work with people. They are finding compassion at the center of almost all solutions, even political ones. They are finally taking responsibility for the creation of their own circumstances, instead of *buying into simply accepting what they are given*. Entire cultures are questioning what they have been told in the past.

The result is all around us, with wisdom replacing greed, monetary systems morphing into something they should have been from the beginning, corporate structure with far greater integrity in the board room. Entire countries are changing the way they have worked an old energy for centuries and the oldest energy on Earth is being seen for what it is—something that stinks—and for the first time, almost everyone can smell it.

Let this book help awaken you to a grander purpose, to find the creator inside and start waking up happy, centered, and hopeful for humanity. If you pick up another book and it does the same thing, then read it and put it on the shelf next to this one. If you find a process that helps teach the principles of this book (even if I didn't write it), then use it! Use the discernment engine that is your own DNA and weigh the energies of what is out there, for there will be much to weigh.

If it sings to your heart, then it's simply a continuation of this teaching. If it helps heal and bring you to a place of understanding, then celebrate it. If you can read it and say, *"It is well with my soul,"* no matter what is going on, then never let it go. If you can weep with joy that you have found your purpose and place in the Universe, then read it over and over. If someone eventually identifies 27 layers, then celebrate them all!

However, if it criticizes, offers dramatic and doomsday predictions, tells you that you have to do something in order to deserve good things, gives dates of your demise then changes them at the last minute, telling you "we saved ourselves," then see it for what it is: Marketing 101.

It's time you knew the truth. As an author, I can tell you that it's easy to sell books and documentaries. Just create something dramatic, dress it up as believable using old outdated predictions of what's going to happen, wrap it with something—anything from Nostradamus—put spooky music behind it, and get that horror specialist announcer guy with emphysema to narrate a special about it. Then enjoy your trips to the bank.

Perhaps, just perhaps, these next years may bring a close to all of this fear mongering? It's up to you, reader. It's up to you.

Blessings to all who have put the energy of their Higher-Selves into the understanding of the principles of this book!

Lee Carroll

Kryon Information

Kryon
Book 12

Lee Carroll

Visit the Kryon E-Magazine

No subscriptions • No passwords
No disclosing personal information
No solicitations for money • Just FREE

FREE ONLINE
Over 70 hours of free MP3
audio channellings!

In the Spirit Epublication

- Kryon questions & answers
- Free Audio and Video
- Photo scrapbook of seminars
- Articles on travel
- Science & technology
- Music & Art
- Healing Techniques
- Laughter & Humor
- Archived articles for reference
- And much, much, more!

www.kryon.com

Drop by and Visit Our
HOME - *page*

The award winning Kryon website allows you to find the latest information on seminars schedules, and Kryon related products. Browse through portions of Kryon books, read some of the most profound Kryon channellings, reference, inspirational and educational material. Read some of the hundreds of answers in the Kryon Q&A section. Enjoy over 70 hours of free channelling audio from all over the world in many languages.

Kryon's Website offers the latest in technology and is easy to navigate. Our main menu allows you to view in an animated or non-animated format allowing for maximum Internet speed.

Find the latest Kryon information at:
www.kryon.com

www.kryon.com

Up Close
with
Kryon

Get together for a personal afternoon or evening with Kryon and Lee Carroll in the comfort of a cozy community center or intimate hotel conference venue with a small group of dedicated Lightworkers. It's the most popular way to join in the Kryon energy in the USA and Canada.

The special meeting starts with an introduction and discussion by Lee Carroll regarding timely New Age topics, then it continues during the day with profound, inspired teachings from the Kryon work. It finishes with a live Kryon channelling. Group size is typically 75 to 100 people. Often lasting up to five and a half hours, it's an event you won't forget!

To sponsor an event like this, please contact the Kryon office: e-mail <kryonmeet@kryon.com>. For a list of upcoming event locations, please see our Website page [www.kryon.com/schedule].

Kryon at the United Nations

Lee Carroll - UN visit 2005

Seven times since 1995, Lee Carroll and Kryon have been invited to lecture and channell at the S.E.A.T. (Society for Enlightenment and Transformation) at the United Nations in New York City. By invitation, he has brought a time of lecture, meditation and channelling to an elite group of U.N. delegates and guests.

Kryon Book Six, *Partnering with God*, carried the first two entire transcripts of what Kryon had to say... some of which has now been validated by the scientific community. Kryon Book Seven, *Letters from Home*, carries the meeting in 1998. The 2005 and 2006 transcriptions are in Kryon Book Eleven, *Lifting the Veil*. All Seven of these transcripts are on the Kryon Website [www.kryon.com/channelling], up through 2009.

Our sincere thanks to **Zehra Boccia** for her help with introducing us to the presidents of this organization over the years. We thank the S.E.A.T for the invitations, and for their spiritual work, which helps to further enlighten our planet.

Index

Accessing the Akash 167
Activation – DNA 23, 73, 75, 126,
.... 184, 222, 229, 231, 236, 242, 276
Akashic Record 24, 53, 114, 124
.... 142, 159, 164, 166, 171, 199, 228
.... 233, 244, 250, 260, 300, 304
Artificial Life 306
Ascension 105, 116, 126
.... 132, 211, 227
Australia Story (Uluru) 61
Barbra Dillenger 33, 85
Big Bang 153, 287
Biology Weakness Puzzle . 178, 265
Cave of Creation.... 142, 165, 173, 304
CERN Large Hadron Collider... 26, 53
Channelling 20, 30, 44, 54, 66, 75
.... 84, 100, 131, 134, 146, 162, 189
.... 223, 259, 280, 283
Charles Fillmore 82
Christ Guide-set story 138
Color Illustrations - DNA 38, 321
DNA - Explained 42, 102
DNA – Activation 23, 73, 75, 126
.... 184, 222, 229, 231, 236, 242, 276
DNA – Artificially Created 306
DNA – Communication 50
DNA – Genes Puzzle 112
DNA – Group One 110
DNA – Group Two 134
DNA – Group Three 152
DNA – Group Four 186
DNA – Imprints 52

DNA – Junk DNA 12, 45
DNA – Layer One 110
DNA – Layer Two 118
DNA – Layer Three 126
DNA – Layer Four & Five 134
DNA – Layer Six 144
DNA – Layer Seven 152
DNA – Layer Eight 164
DNA – Layer Nine 174
DNA – Layer Ten 186
DNA – Layer Eleven 194
DNA – Layer Twelve 202
DNA – Layer Organization 107
DNA – Layers Discussion 70
DNA – Lemurian Layers 67, 152
DNA – Magnetic Field 114
DNA – Merkabah 116
DNA – The Ancients Knew 272
DNA – The History of It 258
DNA – Genome Shotgunning 43
DNA Grouping Summary 214
Donald O'Dell 216
Dr. Todd Ovokaitys 10, 27
Elan Dubro-Cohen 24, 36
Elijah 210, 226
ETs 206
Feminine World Story 195
Gaia Effect 288
Hale Makua 57
Harmonic Convergence 97, 237
.... 275
Hebrew Names 35, 39

Index

Higher-Self 54, 80, 110
.... 122, 136, 142-150, 186, 210
.... 222, 228, 233, 237, 245-255
.... 260, 280, 301, 305, 255
History of the Human Race 258
How the Bible Became 216
Human Genome Project 11, 43, 261
Inner Child 130
Intelligent Design 205, 293, 295
Invisible Stuff 26
J.J. Hurtac 34
Junk DNA 12, 17, 22, 46, 116, 262
Karma 117-127, 120
.... 166, 173, 245, 260
Layer One - DNA 110
Layer Two - DNA 118
Layer Three - DNA 126
Layer Four & Five - DNA 134
Layer Six - DNA 144
Layer Seven - DNA 152
Layer Eight - DNA 164
Layer Nine - DNA 174
Layer Ten - DNA 186
Layer Eleven - DNA 194
Layer Twelve DNA 202
Layers (general information) 70
Lemuria 35, 55, 66, 79, 111, 152
..160-174, 176, 184, 214, 228, 233, 244
.... 247, 254, 261, 263, 269-275, 291
Life Lesson 118
Makali'i 59
Mark of the Beast 219

Matariki 67
Merkabah 116, 171, 189, 227
.... 231, 304
Mining the Akash 168, 250
Nostradamus 96
Numerological Meanings 86
Numerology 20, 78, 80, 83-97
.... 124, 149, 171, 183, 190, 210
.... 219, 243-245
One Kind of Human 55, 268
Parable of the Ancients 262
Phantom DNA Effect 17
Pleiadian-Seven Sisters 54, 56
.... 58-59, 65-67, 153, 156-165
.... 170-176, 184, 194-197, 208
.... 268-275, 307
Pope John Paul 199
Quantum Definition 24
Rosalind Franklin 43
Scientific Bias 280
Superstring Theory 15, 27
Temple of Rejuvenation 177
The "Entourage" of You 140
The Big Scary Secret 215
The Power of Humanity 205
The 72 Names of God 34
Transplants 24, 300
Vladimir Poponin 16, 28
Watson and Crick 10, 43, 50
Woody Vaspra 59
2012 (Twenty-Twelve) 65, 78
.... 197, 208, 272, 274, 310

Free Audio online!

No subscriptions • No passwords
No disclosing personal information
No fuss • No e-mails asked for
... just FREE

[www.kryon.com/freeaudio]

Check out the many channellings and even some former Kryon audio books and discontinued audio cassettes at the above Internet address. Download these free MP3 files and put them in your iPod or just listen on your computer. These are full stereo, high-quality live recordings of some of the most profound Kryon channellings in an updated and growing library. Over 70 hours of listening!

Free MP3 Download	Free MP3 Download	Free MP3 Download
Israel #1	Israel #2	Israel #3
Free MP3 Download	Free MP3 Download	Free MP3 Download
Parables 1	Parables 2	Parables 3

www.kryon.com

The 12 DNA Layers Illustrations
By Elan Dubro Cohen

www.ElanDubroCohen.com

DNA Layer One
Keter Etz Chayim

Kryon meaning:
"The Tree of Life"
The Biological Layer - Double Helix
Part of the "Grounding Group" Layers 1-3

DNA Layer Two
Torah Eser Sphirot

Kryon meaning:
"The Divine Blueprint of Law"
Life Lesson Layer
Part of the "Grounding Group" Layers 1-3

DNA Layer Three
Netzach Merkava Eliyahu

Kryon meaning:
"Ascension & Activation"
Ascension Layer
Part of the "Grounding Group" Layers 1-3

DNA Layers Four & Five
Urim Ve Tumim
Aleph Etz Adonai

Kryon meanings:
"The Light & Power" / "Core Crystal Energy"
Angelic Name Layers
Part of the "Human Divine Group" Layers 4-6

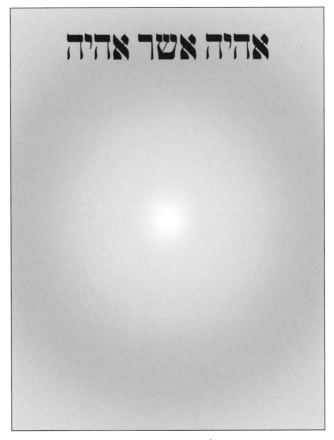

DNA Layer Six
Ehyeh Asher Ehyeh

Kryon meaning:
"I Am that I Am" / "Higher-Self"
Prayer & Communication Layer
Part of the "Human Divine Group" Layers 4-6

DNA Layer Seven
Kadumah Elohim
Lemurian Name: Hoa Yawee Maru
Kryon meaning:
"Revealed Divinity" / "DNA Home Language"
Extra Dimensional Sense Layer
Part of the "Lemurian Group" Layers 7-9

DNA Layer Eight
Rochev Baaravot
Lemurian Name: Akee Yawee Fractus
Kryon meaning:
"Riders of the Light" / "Master Akashic Record"
Wisdom & Responsibility Layer
Part of the "Lemurian Group" Layers 7-9

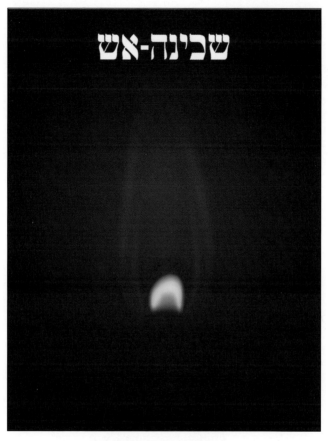

DNA Layer Nine
Shechinah-Esh
Kryon meaning:
"The Flame of Expansion"
"The Violet Flame" of St. Germain
Intelligent Human Healing Layer
Part of the "Lemurian Group" Layers 7-9

DNA Layer Ten
Vayikra

Kryon meaning:
"The Call to Divinity" / "Recognition of God in You"
Divine Source of Existence Layer
Part of the "God Group" Layers 10-12

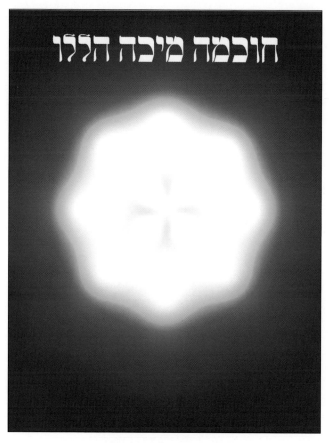

DNA Layer Eleven
Chochmah Micha Halelu

Kryon meaning:
"Wisdom of the Divine Feminine"
Pure Compassion & Mother Layer
Part of the "God Group" Layers 10-12

DNA Layer Twelve
El Shadai

Kryon meaning:
"God" / "Almighty God" / "The God Within"
God Essence Within the Human - God Layer
Part of the "God Group" Layers 10-12

"Let me take you to the interior of DNA itself and reveal to you the esoterics and the love that this process has within it. For more than chemistry, this DNA event defines the core of sacred life, the love of God within the Universe, mixed with dimensional confluences and the joy of creation. DNA is the crossroads of God and man, the mixture of quantum and non-quantum, and it vibrates with the essence of the truth of the Universe."

Kryon

The 12 DNA Layers Illustrations
Beautiful poster size color reproductions!
To Order On-Line:
www.ElanDubroCohen.com
Phone: 1-928-284-3703
Fax: 1-928-284-3704

JAN TOBER is an International speaker, healer, and facilitator. She is co-author of the best selling **Indigo Children** series Hay House books, having introduced the very term Indigo Children along with Lee Carroll in 1999.

Co-creator of the Kryon work, she travels with Lee internationally, bringing her healing voice to thousands.

NEW DNA ACTIVATION PROCESS
with JAN TOBER!

Jan Tober is now offering for the first time, a profound personallized sound activation process. Using the crystal bowls from CRYSTAL TONES www.crystalsingingbowls.com, Jan creates a custom CD recording using information gathered from your personal birth information. This is a DNA activation and ascension process that she has developed over many years, using her healing voice and the renowned quality of the singing crystal bowls.

To Order: **Price $150⁰⁰**. After your initial email contact and details, you will receive your custom CD with a greeting from Jan, and your personal activation. Allow two weeks time. To begin the process, simply email: **jantober@jantober.com** put "DNA activation" in the subject line. A representative will return your email and start the process with you. Credit Cards Accepted

Jan Tober Meditation Products:

As heard internationally at the Kryon Seminars worldwide, available for purchase online at:
www.kryon.com/store

Visit Jan's website at:
www.jantober.com

New!

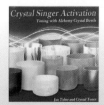

Jan Tober and William Jones combine together for a profound DNA activation album! Using the newest bowls from Crystal Tones, Jan creates almost a full hour of toning and bowls using her voice and the incredible energy of bowls infused with gem stones and minerals that amplify their inherent qualities.

For more information: **www.kryon.com/store**

"DNA is the antenna of the body, listening for the profundities of your epiphanies, the breakthroughs in the tight fabric of fear and frustration. It measures the heights of your joy, the peaks of your passion, and sees the smile on your face as you finally understand that the pathway to God has always been open and available to you. It responds by orchestrating cellular structure to enhance who you are, and to complement your life on Earth. It uses each layer of its own magnificence in the perfect ways that will create the soup of healing, a confluence of love creation, and an honoring of the Human Being's intent. Finally, Human Beings are beginning to understand the Mastery within!"

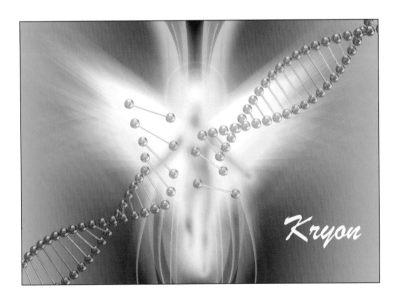